Systems Thinking

Coping with 21st Century Problems

Industrial Innovation Series

Series Editor Adedeji B. Badiru
Department of Systems and Engineering Management
Air Force Institute of Technology (AFIT) − Dayton, Ohio

PUBLISHED TITLES

Computational Economic Analysis for Engineering and Industry
Adedeji B. Badiru and Olufemi A. Omitaomu

Handbook of Industrial and Systems Engineering
Adedeji B. Badiru

Industrial Project Management: Concepts, Tools, and Techniques
Adedeji B. Badiru, Abidemi Badiru, and Adetokunboh Badiru

Systems Thinking: Coping with 21st Century Problems
John Boardman and Brian Sauser

Techonomics: The Theory of Industrial Evolution
H. Lee Martin

FORTHCOMING TITLES

Beyond Lean: Elements of a Successful Implementation
Rupy (Rapinder) Sawhney

Handbook of Military Industrial Engineering
Adedeji B. Badiru and Marlin Thomas

Industrial Control Systems: Mathematical and Statistical Models and Techniques
Adedeji B. Badiru and Olufemi A. Omitaomu

Knowledge Discovery for Sensor Data
Auroop R. Ganguly, João Gama, Olufemi A. Omitaomu,
Mohamed Medhat Gaber and Ranga Raju Vatsavai

Modern Construction: Productive and Lean Practices
Lincoln Harding Forbes

Project Management: Principles and Applications
Adedeji B. Badiru

Process Optimization for Industrial Quality Improvement
Ekepre Charles-Owaba and Adedeji B. Badiru

STEP Project Management: Guide for Science, Technology, and Engineering Projects
Adedeji B. Badiru

Technology Transfer and Commercialization of Environmental Remediation Technology
Mark N. Goltz

Triple C Model of Project Management: Communication, Cooperation, Coordination
Adedeji B. Badiru

Systems Thinking

Coping with 21st Century Problems

John Boardman
Brian Sauser

CRC Press
Taylor & Francis Group
Boca Raton London New York

CRC Press is an imprint of the
Taylor & Francis Group, an **informa** business

CRC Press
Taylor & Francis Group
6000 Broken Sound Parkway NW, Suite 300
Boca Raton, FL 33487-2742

© 2008 by Taylor & Francis Group, LLC
CRC Press is an imprint of Taylor & Francis Group, an Informa business

No claim to original U.S. Government works
Printed in the United States of America on acid-free paper
10 9 8 7 6 5 4

International Standard Book Number-13: 978-1-4200-5491-0 (Hardcover)

This book contains information obtained from authentic and highly regarded sources. Reprinted material is quoted with permission, and sources are indicated. A wide variety of references are listed. Reasonable efforts have been made to publish reliable data and information, but the author and the publisher cannot assume responsibility for the validity of all materials or for the consequences of their use.

No part of this book may be reprinted, reproduced, transmitted, or utilized in any form by any electronic, mechanical, or other means, now known or hereafter invented, including photocopying, microfilming, and recording, or in any information storage or retrieval system, without written permission from the publishers.

For permission to photocopy or use material electronically from this work, please access www.copyright.com (http://www.copyright.com/) or contact the Copyright Clearance Center, Inc. (CCC) 222 Rosewood Drive, Danvers, MA 01923, 978-750-8400. CCC is a not-for-profit organization that provides licenses and registration for a variety of users. For organizations that have been granted a photocopy license by the CCC, a separate system of payment has been arranged.

Trademark Notice: Product or corporate names may be trademarks or registered trademarks, and are used only for identification and explanation without intent to infringe.

Library of Congress Cataloging-in-Publication Data

Boardman, John, 1946-
  Systems thinking : coping with 21st century problems / John Boardman and Brian Sauser.
    p. cm. -- (Industrial innovation ; 4)
  Includes bibliographical references and index.
  ISBN 978-1-4200-5491-0 (alk. paper)
  1. Systems engineering. 2. Problem solving. 3. Systems analysis. I. Sauser, Brian. II. Title. III. Series.

TA168.B594 2008
658.4'03--dc22   2007034222

Visit the Taylor & Francis Web site at
http://www.taylorandfrancis.com

and the CRC Press Web site at
http://www.crcpress.com

Contents

Foreword ix
Acknowledgments xi
The Story So Far xvii

Chapter 1 Perspectives 1
  1.1 If You Leave Me Now 1
  1.2 Areas of Perspective 3
    1.2.1 Or 3
    1.2.2 And 5
    1.2.3 Not 7
    1.2.4 Paradox 9
  1.3 Perspectives on Google 10
  1.4 Returning to Iraq 14
  1.5 Time to Think 16
  Endotes 18

Chapter 2 Concepts 19
  2.1 Just a Thought 19
  2.2 What's the Big Idea? 21
    2.2.1 E Pluribus Unum 21
    2.2.2 The Sameness of Systems 22
  2.3 The Conceptagon 23
    2.3.1 Boundary, Interior, Exterior 23
    2.3.2 Wholes, Parts, Relationships 24
    2.3.3 Inputs, Outputs, Transformations 25
    2.3.4 Structure, Function, Process 26
    2.3.5 Emergence, Hierarchy, Openness 30
    2.3.6 Variety, Parsimony, Harmony 35
    2.3.7 Command, Control, Communications 36
  2.4 Time to Think 40
  Endnotes 42

Chapter 3 Engineering 45
  3.1 Pressed into Action 45
  3.2 The Way We Were 47
    3.2.1 Life Cycles 48
    3.2.2 Passing Through 50
    3.2.3 Eggs Is Eggs 52
    3.2.4 See What I Mean 53
    3.2.5 Spoiled for Choice 55
    3.2.6 Modeling and Simulation 55
    3.2.7 The Long Haul 58
  3.3 Quite Another Story 61
  3.4 Time to Think 63
  Endnotes 64

Chapter 4 Dynamics 65
  4.1 Thinks Can Only Get Better 65
    4.1.1 A Systems Language 65
    4.1.2 Servers and Clients 66
    4.1.3 Archetypes 69
  4.2 Time to Think 75
  Endnotes 78

Chapter 5 Soft 79
  5.1 Breakfast @ Tiffs 'n' Ease 79
  5.2 Softly, as I Lead You 80
    5.2.1 What Seems to Be the Problem? 81
    5.2.2 Getting to the Root of the Problem 84
    5.2.3 Ideally, This Is What We See 86
    5.2.4 It's Good to Talk 89
    5.2.5 The Long Unwinding Road Map 91
  5.3 Time to Think 91
  Endnotes 94

Chapter 6 Systemigrams 95
  6.1 Into Great Issues 95
  6.2 Evolution 99
  6.3 From Prose to Picture 100
  6.4 Going off the Rails 102
  6.5 Normal Service Will Never Be Resumed 107
  6.6 Postlude 111

Contents   vii

    6.7 Time to Think 113
    Endnotes 115

Chapter 7 Togetherness 117
    7.1 Common Causes 117
    7.2 An Intelligent Community? 118
        7.2.1 The Way It Is 118
        7.2.2 The Way It Must Be, OK? 120
        7.2.3 Making Sense of Togetherness: 1 122
    7.3 The Madder of All Wars 124
        7.3.1 Future Capabilities 124
        7.3.2 A New Constellation 126
        7.3.3 Making Sense of Togetherness: 2 130
    7.4 Principles for Togetherness 132
        7.4.1 Coexistence 132
        7.4.2 Cooperation 134
        7.4.3 Coeducation 135
    7.5 Time to Think 136
    Endnotes 141

Chapter 8 Of 143
    8.1 Social Networks 144
    8.2 Order Forms 146
    8.3 Technology Networks 147
    8.4 Less Auto, More Mobile 149
    8.5 The Price and Prize of Togetherness 151
    8.6 The System of Systems Debate 154
    8.7 Essential Characteristics 155
        8.7.1 Autonomy 157
        8.7.2 Belonging 157
        8.7.3 Connectivity 158
        8.7.4 Diversity 159
        8.7.5 Emergence 160
    8.8 Back to Biology 161
    8.9 Time to Think 163
    Endnotes 166

Chapter 9 Paradox 169
    9.1 Make My Joy Complete 169
    9.2 The World of Both 170

9.3 System Paradoxes 172
    9.3.1 Boundary 172
    9.3.2 Control 174
    9.3.3 Diversity 178
9.4 Bothersome Bovines 181
9.5 Time to Think 182
Endnotes 184

Chapter 10 Complex 187
10.1 Life's Rich Tapestry 187
10.2 Sides of Bacon 190
10.3 A Weakness Stronger Than Strength 192
10.4 Ready, Fire, Aim 193
10.5 Snowballs and Seesaws 197
10.6 Significant Others 203
10.7 Postlude 204
10.8 Time to Think 205
Endnotes 207

Index 209

Foreword

It is a great pleasure to introduce this inaugural edition of *Systems Thinking: Coping with 21st Century Problems* by John Boardman and Brian Sauser. In response to the increasing relevance of "systems thinking" to global challenges from terrorism to energy to clean water to healthcare, these authors provide a unique perspective on the word "system." A perspective that causes us to rethink its meaning and rationale, and to reconnect, in a conscious and explicit manner, with the inherent opportunities and difficulties with a "systems approach." This is increasingly necessary for us to address the seemingly intractable systems problems within our society.

The authors first provide the context to systems thinking from an engineering systems point of view, and then extrapolate their discussion to problems that are decidedly societal, where engineering and technology is just an element of an overarching solution. While the authors present pragmatic mechanisms to understand and address co-evolving systems problems and solutions, the primary contribution of this textbook is to initiate critical thinking within the reader while addressing such problems in an attempt to encourage "a systems response." Within an environment where the treatment of subjects such as systems engineering and systems architecting and systems thinking take on a decidedly linear approach, this is a most nonlinear treatment of the subject.

This textbook is ideally suited for business, organizational, and technical leaders as well as political and social leaders. It can serve as a primary text for courses on systems thinking and critical thinking, and as a complementary text for courses on systems engineering and systems architecting. The material in this text represents significant research conducted by the authors in the application of systems engineering and systems thinking principles to engineering systems and enterprise systems, and has benefited from student feedback from multiple courses taught on related subjects by both Boardman and Sauser.

Dinesh Verma, Ph.D.
Stevens Institute of Technology

Acknowledgments

Writing these acknowledgments is itself an exercise in thinking. That is never easy for me, but it is an inescapable challenge. So I decided to use systems thinking as my guide. Who has helped me? When was this? How did it happen? What did they do? Where was I at the time? Why did they bother? Now it is easier. I have begun to think. All I need now are words, the tools of my trade, in the service of grateful thanks to those who have made my world what it is—fascinating, humbling, joyful, an unending learning experience that brings meaning to origin, purpose, and destiny.

In 2004, Dinesh Verma plucked me from the golf courses of central Florida, where I had reduced my handicap from 22 to 14, and planted me in an emerging systems program whose openness and dynamism is helping to reshape industry, academe, government, and, most importantly, individuals. I have not played golf for more than 3 years and do not plan to again until I can confidently make a success of retirement. In the meanwhile, purposeful employment, a principal fruit of which is this book, brings more meaning to me than a birdie. Just about! Thanks, boss.

I attended the University of Liverpool in the early 1960s for three reasons: to discover the secrets behind the Beatles' success, to support the Liverpool Football Club, and to inhabit ale houses. Majoring in electrical engineering was a vehicle to these ends. However, it was the inscription on the university's clock tower atop Brownlow Hill that started to make me think outside the narrow framework that I had created: "This Institution exists for the advancement of knowledge and the ennoblement of life." Wow! I was privileged to meet many people there who lived that out, two I can never forget. They are Brian Swanick and Phil Mars. Scousers from top to bottom. Their intelligence, humor, passion, insights, and candor infected me like nothing I had known or have experienced the like of since. I still believe that if the three of us got it together the world would never be the same. Guys, you have a special place in my heart.

It is people who make the difference in your life. At the time I may have been mistaken about the difference they were making. No one can ever go back, but it is always possible to look back (and not in anger) and gain a better perspective. Accordingly, I owe a huge debt of gratitude to former but unforgotten academic colleagues: Dave Sandoz, Ray Thomas, Sandy Robson,

Ken Woodcock, Barry Wilkinson, Niall Teskey, Graham Wilkinson, Terry Duggan, Graham White, John Pickering, Neil Merritt, Dave Harrison, Stuart Wingrove, Julian Bulbeck, Mike Rose, Matthew Turner, and Ben Clegg. Thanks for the memories.

In the mid-1980s I became involved in several committees of the Institution of Electrical Engineers (now the IET). Committee life has slightly less appeal to me than seeing Liverpool lose a European Cup Final (they have won five to date by the way). But I was thrown together with people of exceptional talent whose energy, intellect, and friendship uniquely restored my confidence and competence to make progress in academe. They are Pete Gawthrop, George Irwin, Pete Fleming, John O'Reilly, Kevin Warwick, Colin Tully, Pat Moore, Andrew Clark, Derek Hitchins, Philip McPherson, and, lastly, Allen Fairbairn—probably the greatest systems thinker never to play for England. All these guys, in my book, are World Cup winners. Thank you for letting me play in the game with you. It was all too brief, yet distinctly memorable.

Industry has always held a special fascination for me. It is the place where mistakes are made, not just thought of. It demarcates the concepts of conclusion and decision. You really do have to be a gambler to play there—someone who is not simply foolhardy but who makes a business of risk. Foolishness I can hack. Calculated risk is something else. Nevertheless, I have been privileged to work in, alongside, and for some wonderful companies whose leaders, in my eyes, have influenced me as an engineer, a person, and, most compellingly, a thinker. Their intellectual impact has been extraordinarily immense, and curiously unsurpassed by professional academics. In my first real job, Denis Farquhar, who ascended to chief engineer of MANWEB, scolded me with the line: "John, you get the colleagues you deserve!" It was intended to abate my complaining attitude. Far too subtle for me at the time. Now I know. In GEC Marconi, I was privileged to encounter Bill Bardo, Bob Wilkinson, Ian Jenkins, and Andrew Farncombe. These guys, metaphorically, could play for Liverpool FC, who incidentally need to sign footballers of that caliber if they are ever going to win the English Premier League. Thank you all for teaching me so much, for showing an interest in my thinking, and especially for supporting my research. In Rolls-Royce, I met many good folk. Hard workers, great firefighters (some of them skilled and undetected arsonists), great engineers, and receptive learners, which is a remarkable attribute in a company with such a world-class reputation for excellence. Chief among this band is Frank Litchfield, to whom I owe a special debt of thanks for teaching me about process and listening to my thoughts on how to leverage it. To all my industry colleagues I render a heartfelt thanks. My systems thinking is what it is largely because of you. I am to blame, but to you goes all the credit.

Latterly I have become acquainted with various elements of U.S. industry, government agencies, and places of higher learning as a consequence first of making this land my home and then my place of work. Consequently,

Acknowledgments

several people have made an impact on my life and my thinking, and to them I extend my grateful thanks: Ruth David, Rick Dove, Steve Krane, George Korfiatis, Mahesh Kumar, David Nowicki, Spiros Pallas, Art Pyster, Donna Rhodes, Jack Ring, Mark Schaeffer, Todd Tangert, and Mark Wilson. Thank you all for being in this brave new world. With your cooperation I look forward to many years of activity that will be fascinating and fun.

Writing a book is easy. You simply have an idea, sling a few words together, and press on. When it is complete, you get some folks to read it, to say what they think, ask them if it makes sense, and how it might be improved, never believing this is actually possible. This is when thoughts and realities collide. It would be great if readers adored all the things that a writer enjoyed at the time the words hit the page. Why they do not amazes me. No! It disappoints me. Then I have to think again. To practice what I preach. To listen and to learn, which is what I would have my readers do. I am a better man for this. Not at the time and never immediately. But finally. Brian Sauser and I have had the benefit of many willing and beautifully capable reviewers. Intelligent, respectful, experienced people—students, faculty, and colleagues—who have wanted to be part of this book. We truly owe them an extended amount of sincere thanks. They are Mark Weitekamp, John Wirsbinski, Robert Edson, Michael DiMario, Steve Bishop, Clif Baldwin, Larry John, Nicole Long, Thomas Ford, Alex Gorod, Donny Blair, Vishwajeet Kulshreshtha, and Devanandham Henry. Thanks guys. Your inputs have transformed our work, making our outputs more than we ourselves could have achieved. And while writing is one thing, actually producing this book is another. For that, we express our utmost appreciation to Taylor & Francis, including Cindy Renee Carelli, senior acquisitions editor, Jill Jurgensen, production coordinator, Jay Margolis, project editor, Eva Neumann, typesetter, and Nadja English, senior marketing manager.

This book is in part a response to the creation of the systems thinking course that forms part of the graduate program in the School of Systems and Enterprises at Stevens. Over the past 3 years almost 200 students have taken that course. Whether they know it or not, each one has impacted this book. Some will be able to find their thoughts, maybe even their words, in these pages. As much as any book can be living, this is, because it is an organic part of a greater whole that includes this student body. May that body grow and become wise. May this book help that process. Thank you, you great unknown. We know you, and we even remember some of your favorite movies!

Speaking of organic wholes, authors themselves are part of a larger mix. In our case that is Stevens and, in particular, the School of Systems and Enterprises, which has emerged as a consequence of this incredible dynamic that feeds on the desire for education, structure, progress, comprehension, and systems of all kinds—technological, linguistic, computational, organizational, and entrepreneurial. We owe undying gratitude to our Stevens colleagues, some of whom must of necessity be singled out for their loyal support, inspirational interventions, and critical thinking. They are, first,

Elaine Chichizola, Fionnuala Coyle, Shobi Sivadasan, and Kathy Connors. Ladies, you are priceless in our eyes. Thank you hugely for your work and comradeship. Second is Mike Pennotti, who not only manages the graduate program but also manages to think and hugely influence our thinking. Third is Ralph Giffin, who keeps order seemingly without ordering anybody to do anything, and who puts his walk about leadership in line with his talk about it. Ralph, thank you for getting me to read again. Finally, there is that quite remarkable young man, a unique talent who is undiscernibly changing the world, one act of random kindness at a time. Dinesh Verma has been willing to be the guy "they will hang" by assuming the mantle of dean. I do not think they will find the rope, pal. There are far too many of us to keep it hidden from them. Thanks for being you and making this book happen for us.

Finally, a few personal remarks. First, my coauthor. I have been around a long time now and known a lot of people and done a few things. Never in all this time have I come across anyone like Brian Sauser. I wish I knew what it was about this boy. Maybe it is not about him. Maybe it is about me and that long time, filled with foolishness and errors and brilliant ideas and laughs and tears. Whatever it is about, Brian Sauser is to me the most special person outside of my family. To the point he is family. Brian, it is a privilege to know you and an unsurpassed joy to work with you. May we work together for a long, long time. I am hoping forever.

My mother, Charlotte, put shoes on my feet, clothes on my body, food in my mouth, ideas in my head, and hopes in my heart. Footwear, flannels, and feeding has frequently come and gone since those first days. But the desire to imagine, to believe, to aspire, and to inspire endures. For all these gifts, wrapped in God's love from a devoted mom, I shall be forever thankful. Thanks mother.

My children, Richard John, Sarah Elizabeth Renee, and Jonathan Paul, are individually and together immeasurable blessings to me. They continually surprise, delight, thrill, and inspire me. I love you. I know you know what this book is really all about because you live its contents out so imaginatively and practically every single day. God bless you.

Lastly, my wife, Alison, is a special person. A truly treasured possession. Her love, faithfulness, and companionship are as clear a view of heaven that I could ever have. Thank you for giving me up to my study for seemingly endless hours, for tolerating my irascible behavior when it has not gone well, and for blissfully accompanying me on flights of fancy when it has been ecstasy. I love you, with and without thinking.

John Boardman
Blairsville, Georgia

I began my systems journey not as an engineer, but as an agriculturalist! For me, like von Bertalanffy, the understanding of systems is in our understanding of biology.

My B.S. in agriculture development from Texas A&M University showed me a grassroots appreciation for both plant and animal systems based on a rich heritage in agriculture. For this I thank Malcolm Drew, Alvin Larke, Jr., Christine Townsend, and all my mates from the Mule Barn … Whoop! My M.S. in bioresource engineering from Rutgers, the State University of New Jersey, showed me that biology and engineering could two-step the fragile edge of chaos and order. For those that I danced in the light of their candle, I thank Gene Giacomelli and K. C. Ting. As for those who walked beside me and spent many nights in philosophical discussion about life (and thus systems), I thank Ann Marie Carlton, Dave Fleisher, and Luis Rodiriguez. Finally, my Ph.D. in technology management from Stevens Institute of Technology allowed me to understand the responsibility and accountability people play in the governance of biological and engineered systems. I attribute this to George Korfiatis, Richard Reilly, Peter Koen, and Ed Hoffman, but my fortitude in academics will forever by attributed to the guidance of Aaron Shenhar, my advisor, colleague, and friend.

My professional career has taken me from government to academia to government and back to academia. I started my career managing an applied research and development laboratory in life sciences and engineering at NASA Johnson Space Center. It was there that people like Don Henninger and Dan Barta taught me about systems design and optimization, and that defining the systems boundary is the hardest thing for a systems engineer to do. From there I was the program director of the New Jersey–NASA Specialized Center of Research and Training where I learned that people are as much a part of the success of any system as the technology is, and I owe all my gratitude to Harry Janes. He was a systems thinker by shear passion and genes. In my last stop before coming to Stevens, I was at NASA Kennedy Space Center developing commercial partnerships, where my dear friend and colleague Joe Robles always challenged my way of thinking and held me accountable for what I said, and still does. He once said that I approached every problem with a systems perspective, to which I replied, "That is because life is a system." And finally, as a professor at Stevens Institute of Technology in the School of Systems and Enterprise. For this, I thank the Stevens community and culture; no plant can grow without the right environment. Throughout this journey, one thing has remained true: I have been guided by systems shepherds.

And as this book goes to print, I stand at the edge of the future with the universe as my systems boundary and standing beside me is what I believe will be the greatest of my systems shepherds, John Boardman. John, we will not know what we are to be … the system will tell us. You will never truly understand the magnitude of your influence … and to understand is to not know. Thank you!

Finally, I thank my family: you gave me love and support as I progressed through the systems journey of penning this book. My parents are the locus of my being and the foundation of my thinking—I will forever be in your gratitude. My wife, Meg, is everything I dream and always keeps me grounded in reality; without her I would be less of a man. Meg, thanks for tolerating me with love. My vibrant, loving, and beautiful children, Gabriella and Hunter, you make my life complete.

So, let the journey begin …

Brian Sauser
Somerset, New Jersey

The Story So Far

Albert Einstein reportedly said, "In the brain, thinking is doing." Now perhaps his remark was offered as a counterpoint to the prevailing argument that doing is more important than thinking. As engineers, we both realize all too well that no matter how much you think, unless you do, "it ain't gonna happen!" You can plan, prepare, and predict all you want, but action occurs through doing, and that is what matters. Of course, many engineers frustrate the heck out of a lot of people by getting things done, and getting them done remarkably well, without apparently much attention being given to the thinking behind the doing. Philip Sporn once eloquently remarked:

> The engineer must often go beyond the limits of science, or question judgment based on alleged existing science. He must frequently assert his own overriding judgment and stake his reputation to go into areas beyond that which has been fully explored scientifically, and indeed, may even contravene that which has been claimed to have been demonstrated incontrovertibly by the science of the day.[1]

It is as if the engineer makes thinking happen by simply doing.

There is a value in doing per se, a value that thinking can never lay claim to, and regrettably a value that is inordinately esteemed to the very detriment of thinking itself. We are obliged to Dr. Einstein for helping to redress this imbalance. But perhaps the renowned scientist was merely being beguiling. Maybe he wanted us to see that though we make this evident distinction between thinking and doing, at some level of observation there is none. What is the thinking of a neuron whose dancing with neighbors produces a thought that transcends the totality of neural choreography?

This book is about thinking, and it describes specifically a type of thinking that makes sense of togetherness by deploying the notion of *system*, a term that has achieved, in a relatively short space of time, unparalleled ubiquity. Our engineering friends believe the term *system* is theirs of right and they alone understand systems. After all, who builds them? Who gets the job done? You would think, to hear some engineers talk, that they invented the term itself. In fact, what propelled it into the high-currency values it occupies today were the ideas of Ludwig von Bertalanffy,[2] an Austrian scientist who, in making sense of plant life, wondered whether his line of reasoning, essentially systemic, could have application to many other forms of life, including the societal and the technological. Could plants, animals, rock formations, persons, peoples, and the products that knit us together, he reflected, be regarded as systems? It is thought like this that this book values and commends.

From an Austrian émigré to the grandson of Italian immigrants. Rudolph Giuliani's book on leadership is replete with references to systems. Here are some:[3]

- "For any system to remain effective it must continually challenge itself."
- "Despite the failures of some very high profile American businesses and the alleged 'corporate greed' that caused them, the reality is that they reflect only a small percentage of the business community. And all of these collapses actually demonstrate that America's economic system is a very healthy one. Furthermore the system is self-correcting ... accountability works to improve all systems."
- "In a system as complex as New York City ..."

Interpreting Mr. Giuliani's thoughts, we believe he envisages at least three kinds of systems:

- The information management kind that gathers and processes data and makes reports that inform action to improve what is monitored
- The task forces themselves, who use these information management systems in order to stay on top of the job of managing a complex community of people, ensuring its well-being
- The cities, states, nations, and their infrastructure that holds these entities together, and adaptively so

Giuliani's book presents many principles of leadership for us to digest, but the telling of these relies, somewhat imperceptibly, on the more fundamental notions of systems thinking.

Einstein also knew the value of system as a notion to stimulate thinking. Here is another of his many remarkable utterances that gives evidence of this: "Gravity doesn't explain two people falling in love." His point here is surely that there exist multiple perspectives, differing levels of behavior, and that there is a real need for different (systems of) knowledge, interlinked for sure but each emergent in their own right, which we need in order to deal with the vastly differing phenomena in our world. These systems of knowledge are the sciences—physics, chemistry, biology, physiology, psychology, and so on.

Einstein, a physicist, was a systems person. So too was von Bertalanffy, who was a biologist. And now we have Rudy, an aspiring 44th president of the United States, perhaps, but by profession an attorney. Evidently, engineers are not the only professionals occupying the systems landscape. And that is a good thing. Of course getting these various disciplines to come together on this landscape, in order to better understand this term *system* and make sense of togetherness itself, is quite another thing. It is a challenge that this book does not duck.

We are suggesting that systems thinking can be thought of in two ways. First, and the obvious one, is to think *about* systems—in other words, to use our mental capacities and the tools we have acquired for cognizing, analyzing, and synthesizing to ruminate on the systems that confront us. In this book we provide a conceptual framework, a system of concepts, in fact, for focusing one's thinking about such systems, whatever shape and size they come in. This is the subject of Chapter 1.

We also describe concepts, advanced by engineers and systems analysts, to help organize one's thoughts and actions relative to the systems of interest, and specifically to their design. This is the subject of Chapter 2.

Because our primary constituency is the engineering profession, we want to draw out the major contributions that this human endeavor has made toward thinking about systems. Engineering systems—in the broadest sense of the term as, for example, MIT has determined it—is the subject of Chapter 3. In presenting this effort, we want to not only provide an assessment of the tactical maneuvers that engineers make, using case studies to illustrate, but also present our essential understanding of the endeavor—in other words, what is the essence of systems engineering. In that way we can present engineering systems as a methodology for dealing with a certain class of system and its affiliated problematique, which then can be compared with other methodologies concerned with squidgier issues, which interestingly are increasingly having an effect in the lives of many systems engineers.

To conclude our survey of ways and means to think about systems, we present in Chapter 4 a summary of the work of industry dynamicists and their specific contribution of a language to describe systems in action, not only in being, and how cause-and-effect relationships play out in continuous dynamical interactions, captured in elegant causal loop diagrams. These four chapters, as a whole, encompass our thoughts on how to think about systems.

A second and crucial way to exhibit systems thinking is to think *from* systems. In other words, to use systems, captured in diagrammatic form somewhat similar to causal loop diagrams, in order to focus our thinking on the very issues that gave rise to our need to think, and subsequently to act. So thinking *about* systems means making systems the focus of our thinking, and our thinking tool kit provides the lenses that constrain and shape our thinking, which might otherwise be chaotic. Thinking *from* systems means to use systems, more correctly systemic descriptions of a problematique and any accompanying treatments of this, as the lenses, with the issues—and issuers—as the focus of our thinking. This approach raises many new considerations that have hitherto been overlooked or deliberately marginalized in favor of executive action. Remember, without doing it, ain't gonna happen! However, there is increasing concern that problem solving, in the guise of providing bigger and better solutions, that is, systems, translates into problem moving, or worse, problem creation; thus, the very considerations that have been overlooked must now be brought onto center stage in order to eradicate this nugatory action.

In Chapter 5 we rehearse the soft systems methodology, which is an attempt to use systems thinking to create conceptual models from ideal perspectives of systems that could conceivably exist and then use these systemic models of ideality as a basis for exploring what actions can realistically be taken in order to ameliorate the circumstances that gave rise to the investigation in the first place.

In Chapter 6 we introduce our own invention of a certain type of conceptual model, the *systemigram*—a portmanteau word derived from systemic diagram. We show the source of our original thinking, how and why these devices have proved useful in dealing with "wicked" problems,[4] and we detail for the benefit of the reader how to create and deploy these with confidence and reasonable expectation of a resolution of conflicts. In Chapter 7, we apply systemigrams to bring context to the meaning of togetherness—one that impacts their structure, behavior, and realization, for the distinction comes from the manner in which parts and relationships are gathered together and therefore in the nature of the emergent whole.

In Chapter 8 we address a contemporary phenomenon in the systems world, *system of systems* (SoS), and we do so by providing our own ideas of how the SoS might differ from an ordinary system, and therefore how they might be differently cognized, analyzed, and synthesized. Unsurprisingly, we suggest that both forms of systems thinking, *about* and *from*, are needful to explore the SoS phenomenon.

We discuss variations of SoS from the traditional technology SoS in which complex and preexisting technology-based systems must be made to work together in order to fulfill some higher-level capabilities that were originally (i.e., at the time of their original design) unforeseen to the problem-solution coexistence SoS whereby we explicitly and purposefully portray both problem-system and solution-system side by side. For wicked problems this pair dances endlessly, rhythmically, and inventively, with improvisation being the hallmark of the choreography. It is what we call systems jazz. It is the uncharted territory on the systems journey. Curiously, the capabilities of the SoS may even yet remain unforeseen and indeed unforeseeable. But because we are creating an SoS, as distinct from an ordinary system, this matter of the unknowable unknowns poses less of a problem. We also discuss the people-rich SoS that has been called the *extended enterprise*, meaning a system of autonomous organizations that have bought into the need to cooperate and collaborate, for individual and collective good.

The Story So Far

Our final two chapters are a deliberate betrayal of all we have said thus far. We hope you are suitably intrigued! Our entire lives have been conditioned by choices and the determined free will to make a decision, to make a choice. How else can you move forward, take action, make things happen, without making a choice? It all makes sense. Yet we are convinced that we are now confronted with a world where it is less about *or* and more about *both*.

We live in a world of multiple perspectives, myriad stakeholders, competing technologies, diverse partners, and tough competitors. We make sense of this togetherness by choosing one: a single perspective, a major stakeholder, the right technology, a lifetime partner, or the most threatening competitor. And that is all okay, because it is what we do. In this book we want to present an alternative, ironically! We want to suggest confronting this multiplicity by believing all. The risk of believing all is to believe none. We take that risk. If quantum theory is correct, and quantum computing is betting on this, then *both* is possible. The particle is at one energy level and another, apparently instantaneously. The binary digit is both 1 and 0, a qubit. For years, according to our knowledge, the electron has been both particle and wave. And for some, God has been three persons in One.

We chose to make our opening chapter about what we regard as the cornerstone of systems thinking—perspectives, their multiplicity and tenability. It is always great when these can be accommodated simultaneously. But if they cannot, what then? In Chapter 9 we explore the role of paradox in systems thinking. Paradox is the province of conflicting perspectives in which each is true and both cannot be true. It is the harshest form of simultaneity, unless you accept it and therein find the way out. We show how much more evident paradox is, especially in engineering, than we realize, and we provide some principles and a framework for dealing with paradox.

In our final chapter, we explore another duality that is an apparent contradiction—chaos and order. By examining the meaning of complexity and the phenomena that have been observed, mostly in the natural and life sciences, which so beautifully and mysteriously portray this duality, we find more evidence of paradox and clues for dealing with it. Of course, being men of action we are not content to leave complexity to the theoretician and make it a matter of mere observation. In other words, we cannot settle for thought without action. Therefore, we go on to show how complexity thinking has some fruitful applications in the engineering world, provided of course that the mindset of choice does not blur the vision of uncomfortable togetherness that paradox and complexity bring.

For us, Dr. Einstein's remark about thinking and doing is most especially welcome because it points to two essential system thoughts—those of separability and integrability. In regard to thinking and doing, of course you have both. But in the brain they are indistinct. In regard to systems thinking and systems practice, you have a second both: separation and integration. The systems person, practitioner and theoretician, can see the parts, thereby esteeming separability, and the whole, thence valuing integrability. He or she can see both at the same time. That mindset and that skill are something we desperately seek to communicate from this book. In regard to this last point, we have alluded to the parts of this book, in the guise of its chapters. In so doing we vaunt separability and distinctiveness. But we have also argued for the integrability of these parts, for example, in the ways of systems thinking, thereby forming a whole that is hopefully an enjoyable and informative experience. This is something for you to think about. But not unless you do something. Enjoy!

Endnotes

1. Sporn, P., *Foundations of Engineering*, Pergamon, New York, 1964.
2. von Bertalanffy, L., *General Systems Theory: Foundations, Development, Applications*, George Braziller, New York, 1969.
3. Giuliani, R., *Leadership*, Miramax Books, New York, 2002.
4. Rittel, H., and M. Webber, "Dilemmas in a General Theory of Planning," *Policy Sci.*, 4, 155–69, 1973.

chapter one

Perspectives

1.1 If You Leave Me Now

Should the United States pull out of Iraq? Or should we stay? On April 9, 2003, Saddam Hussein's government lost control of Iraq as U.S. forces advanced into the center of its capital. In a symbolic moment, U.S. soldiers helped a cheering crowd of Iraqi citizens pull down a giant statue of the ousted president. After only a few weeks of fighting following the first missile attacks on the city it seemed as though the war was settled and over. On May 1, President Bush formally announced an end to the major combat operations, declaring that the United States had prevailed in Iraq. However, he warned that "difficult work" lay ahead.

Less than 3 months later the head of U.S. military operations acknowledged that attacks on the occupying U.S. troops bore the hallmarks of a classic guerrilla-type campaign. On August 19, a huge bomb demolished the United Nations headquarters in Baghdad, killing twenty people. It was a suicide attack, and a number of international agencies decided to pull their staff out of the capital. November 2 saw a Chinook helicopter shot down by insurgents, killing fifteen U.S. soldiers and wounding twenty-one more. On December 13, Saddam Hussein was found in a cellar of a farmhouse near his hometown of Tikrit by U.S. soldiers. It is an early Christmas present, bringing hope of a timely end to the insurgence, violence, and disruptions to a process of peaceful transition to Iraqi democracy. It is a false hope. In 2004, 849 U.S. soldiers are killed and a further 8,002 wounded. The year is long and bloody. The targets widen. On March 2, 180 Iraqi Shias are massacred in the cities of Karbala and Baghdad as they seek to celebrate the climax of a holy ritual.

The following year gets off to an ominous beginning. On January 20, an audiotape apparently made by one of the insurgency leaders, Abu Masab al-Zarqawi, warns the fight against U.S.-led forces could continue for years. The tape, posted on an Islamist Web site, denounces Shia Muslims for fighting alongside U.S. troops in Iraq. A report issued in July 2005 estimates more than twenty-five thousand Iraqi citizens have been killed since the start of the war in 2003. Based on media reports, the dossier attributes one-third of these deaths to U.S.-led forces. One month later, one thousand Shia pilgrims are killed in a stampede in northern Baghdad. Panic spreads alongside the rumors of suicide bombers in their midst. It is the largest loss of life in a single incident since the invasion. December 15 saw a huge turnout for voting in the elections across the country for full-term government. It is Christmastime and hope returns.

But on February 22, 2006, two men blow up the golden dome of the al-Askari shrine in Samarra, one of the holiest sites in Shia Islam. The year is marked by the execution by hanging of former president Saddam Hussein, found guilty of crimes against humanity. But the death toll for U.S. soldiers has passed three thousand and the Baker-Hamilton report is published recommending a change in policy.[1] The makeup of Congress has changed following elections in November. The Democrats control the Senate and the House and President Bush finds the war in Iraq is now being fought out on Capitol Hill. Those first shots fired in early 2003 are now ringing around the White House itself.

It is 2007 and the American people are war weary. They want their men and women in uniform to come home, their blood no longer to be shed, and American treasure to be directed to better ends. What is your perspective?

The question of Iraq is an emotive one. People feel strongly about it and differently as to what to do. When emotions run high, it is at the risk that intelligence will drain away. So much bloodshed, so little progress. Such a strategic location. Think of the oil, think of democracy, think of freedom. Think of lives lost, think of contracts gained. Think of the future of America and stability in the world. Think of war as the continuance of negotiation by other means. Think of the dead, think of sacrifice. Think of terror. Think.

Systems thinking is a deliberate attempt to think when thinking itself is put at risk by emotion, confusion, and confrontation. When the thinking process is being assailed and overwhelmed by debate, opinion, doctrine, and information, systems thinking stands in the breach and says, "I can help."

Systems thinking does not suppress or supplant perspectives; it adopts them and finds sense in their multiplicity and diversity, their surprise. It does not guarantee success, but it does make thinking possible when that seems impossible. Systems thinking asks, What systems are involved here? And when we find those answers to be unhelpful, it asks, What kinds of systems might be involved here that we have not thought about before? To answer that question, systems thinking has to come up with some new ideas, concepts, techniques, and tools. Our book presents the choicest of these and illustrates their application to problems as large as Iraq and as familiar as children's puzzles.

We begin with taking a closer look at perspectives, at what they mean and what they might mean. We look at four distinct but interrelated areas in which perspectives arise. First, we have the area of *choice*, where there is a decision to be made, a route to take, and of all those choices that compete for our favor, only one can be made. The choice may be clear or unclear, and multiple perspectives are offered to support one choice or another. Second, there is the area of *both*, in which every single perspective is valid and each must be treated with due respect (and diligence). It is not a case of which of these is correct, since quite conceivably the candidates may be opposed,

Chapter one: Perspectives 3

but rather a question of "What is the correctness of the simultaneity of all these perspectives?" Third, we look at the odd case of the inverse whereby if one perspective is volunteered, an opposite immediately arises. This may be the case where people who are continually diametrically opposed will look for an opposite purely to continue their antithetical behavior. Or it may be an important dialectic to exercise whereby suspension of belief in our own correctness helps us remove any unseen or unfelt blinders and see things differently. Lastly, there is the area of paradox to which we later devote an entire chapter. In this case, perspectives both support and oppose one another, making it peculiarly difficult not to dismiss them out of hand for being deliberately frustrating and distracting. And yet, paradox holds an intrigue beyond which lies the hope of greater wisdom. It is helpful but increasingly rare when situations arise that fall neatly into one of these categories, except perhaps the paradoxical one. But the complex issues, and systems, we encounter today, and more so in the future, do not fall neatly into one category. They are an intricate interweaving of all of them, at many different levels and with multiple facets. But just as electronics engineers know how to build any digital logic circuit from a few simple gates—*or*, *and*, and *not*[2]—so we suggest that complex situations can be accessed by these areas of perspective.

1.2 Areas of Perspective

1.2.1 Or

The Mississippi River is one of the world's major river systems in size, habitat diversity, and biological productivity. It is the longest and largest river in North America, flowing 2,315 miles (3,705 kilometers) from its source at Lake Itasca in the Minnesota North Woods, through the mid-continental United States, the Gulf of Mexico Coastal Plain, and its subtropical Louisiana Delta. *Mississippi* is an Ojibwa (Chippewa) Indian word meaning "great river" or "gathering of waters"—an appropriate name because the river basin, or watershed, extends from the Allegheny Mountains in the eastern United States to the Rocky Mountains, including all or parts of thirty-one states and two Canadian provinces. The river basin measures 1.81 million square miles, covering about 40% of the United States and about one-eighth of North America. Of the world's rivers, the Mississippi ranks third in length, second in watershed area, and fifth in average discharge.

The Mississippi River and its adjacent forests and wetlands provide important habitat for fish and wildlife and include the largest continuous system of wetlands in North America. The river supports a diverse array of wetland, open-water, and floodplain habitats, including extensive habitats on national wildlife refuges. Yet human activities have greatly altered this river ecosystem. Most of the river and its floodplain (defined as the adjacent, generally flat surface that is periodically inundated by floodwaters overflowing a river's natural banks) have been extensively modified for commercial navigation and other human developments. Much of the watershed is intensively cultivated, and many tributaries deliver substantial amounts of sediment, nutrients, and pesticides into the river. Pollutants also enter the river from metropolitan and industrial areas. Some of the great U.S. cities that lie along the river are Minneapolis, Memphis, and New Orleans.

Now here is the point—how do two communities separated by such a river form a connection? This is a challenge, a puzzle really, typically thrown at a class of students to exercise their ingenuity while exhibiting a degree of realism bounded by their experience, expertise, and enthusiasm. Of course, it is that very experience that they call upon to nominate candidates, for example, a bridge or a tunnel. Others that we have heard in our time include ferries and hot air balloons. No one has seriously pursued the choice of human cannonballs. One surprising nomination was "divert the river." While this choice is not in the same league as the suggestion that it is better to lower the Atlantic than to raise the Titanic, it does betray a trait of passing the problem onto someone else as opposed to dealing with the one that you face. However, whatever candidate is chosen, the connection of the communities will produce new problems for both—that is the nature of systems. While many candidates can be identified, only one of them is to be pursued. The problem is deliberately framed as an *or* perspective even though in principle a multiplicity of connections can be supported and inevitably are, for example, the Ben Franklin and Walt Whitman bridges that span the Delaware joining Philadelphia with its New Jersey neighbors. Our focus is on what thinking is called for in addressing the *or* perspective.

One thing we have to think about is to be able to generate the choices or candidates. No one really knows where an idea comes from let alone what makes a good or original idea, but we do know that they appear. Some ideas come from past experience—it is just a matter of memory retrieval. But where do the new ideas come from? How can you think of things that you never thought of before? Clearly, a second thing to think about is how to select a candidate from the many that compete for our attention. Engineers love to talk about metrics and criteria and trade-off studies, and this is all very useful equipment that they bring to their domain and from which other disciplines, given some abstract thought, can possibly make good use of also. These two things—identification and selection—seem to be prime foci for whatever thinking we do. A third focus, sadly too often overlooked, is that of considering the need rather than merely accepting it as a given. In other words, what are the needs of these communities to be connected? What drives this? How persistent is it? Can it be expressed differently so that the overriding notion of physical movement between the two communities is not the only influence on the identification process? Also, what are the constraints these communities put on the satisfaction of that need? These are not only applicable to the selection from a list of candidates, but they are also influential on the generation of the candidates' list. So if someone can say

Chapter one: Perspectives

"Let's divert the river," which effectively forms a single community since the interposing dry land can now become common ground, cannot another say "Move community B to the other side of the river and join community A in a new adventure"?

These then are three foci for our thought process: profiling the need, identifying the candidates, and adopting a selection mechanism. And for systems thinkers these thought foci *interact*. They are parts of a wider system. They have relationships, such that changes in any one can affect others, and in ways that might cause the relationships between them to change. And it is the whole of these foci, whatever that might be, that constitutes the system of thinking. And we are not speaking of only the instantiation of this system for a given example—the need for two communities separated by a river to have connectivity. We are interested also in the exploration of this system of thinking at an abstract level so that we might be better prepared for the particular problem itself when it comes along, sometimes in ways that are suggestive of past *or* perspectives and sometimes not. That is how the *or* perspective itself becomes better known or more accessible to refinement.

1.2.2   And

We both played sports as kids. Two different types of football at least. The real deal (soccer) and the American version of football. We also remember those coaches' talks at halftime when our teams were losing badly and something needed to be done. Great coaches diagnose the problem and know what the solution is—maybe a change in tactics, or a key substitution, or a change in roles for some players, or more aggression, or singling out one or two key players from the opposition and putting them out of the game (safely, that is). Whatever it is, you do not know if it works until the final whistle, and by then it is too late. But whatever our differing experiences we do agree on this point—great coaches listen to their players; they get them to talk and to say what they think is wrong, what changes need to be made and why, and how to put those changes into effect. All of this takes place in the heat of the all too brief interlude with passion, when people are frustrated and maybe hurting from bruises, and a sense that everything is at stake. The value of listening is that people give vent to their emotions, they are heard and respected, and they hear one another. Whatever comes out of that mix, there is buy-in and a renewed commitment to do things differently, to make change work, and to refocus on the one goal that all members of the team share: victory.

What is true for sports, including games of sheer inconsequence, is also true of government cabinets. Is it halftime in Iraq or are we close to the end? What does the cabinet think? More precisely, what does each member of the cabinet think? What are they telling the president? And what sense does the president make of this collective wisdom. What buck does he get?

From changing room to cabinet room to chamber room. Different venues, same situation. Four blindfolded men are led into a chamber room to sense one single object. They cannot see what the object is, but they can touch it, smell it, and hear it. They occupy different parts of the room and sense different parts of the same object. After a little while they try to make sense of what they discover. The first person concludes the object to be a spear—having felt something long, hard, and with a sharp point. A second person concludes differently. He senses a snake. True it is long, but it is not rigid; it is flexible and swishes around. He thinks the sound he hears is that of a hissing, but maybe it is the swift movement of the object itself. Not a spear, not at all. The third man differs, yet again. He believes the object to be a fan. Where he is stood there is an unmistakable draft and the drumbeat noise he hears is surely that of an oscillating mechanism, a fan. Lastly, the fourth companion makes his judgment known. It is none of the above. It is a pillar. It is tall, circular, and solid. A sturdy object, and maybe there are three more, one at each corner, the collection clearly capable of bearing a huge weight collectively. Four reports, zero eyewitnesses, but each given with accuracy and confidence with conclusions that could not match less. Who or what is in the chamber room?

Some readers may know the "answer" to this little puzzle. We congratulate them, but we further challenge them to change their mind, to come up with something new, to not be satisfied by simple memory retrievals but to learn what the puzzle is really all about. Others may not know. We want to help them find an answer. Equally, we would like to be able to help President Bush or the team losing at halftime to come through at the final whistle. The way we know to do this is to find the *thought foci* for the *and* perspective.

The prior existence of the thought foci for the *or* perspective is a help here. With these we settled on three entities: candidates, need or objective reality, and choice or discriminating mechanism. The candidates had to be generated or conceptualized. The need or objective had to be validated to ensure that the work in generating candidates did not prove nugatory. And the discriminating mechanism had to be developed, something by which the candidates could be prioritized or differentiated in order to select an optimum. These entities have reuse value in the *and* perspective. But the verbs that apply to them change.

There is no shortage of preexisting candidates. There are at least as many as there are sensors (in the case of the chamber room), advisors (in the case of the cabinet room), or players (in the case of the changing room). Sensors, advisors, or players—one might refer to these as types of stakeholder, an entity that has a valid role to play. So it is not a case of generating these but rather validating them. Does what the sensors conclude make sense? Is what an advisor has to offer, essentially a subjective assessment, of value in the mix? Can what a player brings to turning the game around work, given what others think? Our first thought focus for the *and* perspective then is the validation of subjective statements.

In the same way, the prior need or objective reality does not exist for the *and* perspective, whereas of course it did and was to the fore of thinking with

Chapter one: Perspectives 7

the *or* perspective. This leads us to conclude that there is a thought focus to deal with the generation or synthesis of a need or objective reality based upon the validated subjective viewpoints or candidates available a priori.

Finally, there is a thought focus that provides an integrating mechanism whereby the validated candidates can be synthesized into a single objective reality or need that singularly makes sense of the variety of viewpoints available. By contrast, the *or* perspective required a differentiating mechanism to distinguish the candidates that allowed some prioritization or filtering of the less attractive from the optimum.

In summary, therefore, for the *or* perspective the foci are generate candidates, develop differentiation mechanism, and validate the need. For the *and* perspective the foci are validate candidates, generate the need, and develop an integration mechanism. If that helps, go ahead and motivate the football players, assist the embattled President Bush, and identify the mystery object in the chamber room. But if you need more guidance, and we suspect you will, this book is for you.

1.2.3 Not

We are all quite familiar with contrary behavior. Sheep that will not be herded, subordinates who will not conform, infants who refuse to eat their food. It is as if the very thing you want these miscreants to do they stubbornly refuse and do the exact opposite. Sometimes you have to offer them the opposite of what you want them to do in order to get them in line with your true goal. Lenz's law is a beautiful illustration of this perversity. Insert a bar magnet into a conducting cylinder wrapped in a coil of wire and this will induce an electric current in the wire that then produces an electromagnetic force that opposes the bar's insertion. On the other hand, withdraw the bar magnet and a second current is induced, flowing in the opposite direction, creating a second induced force that resists the bar's withdrawal. This laissez-faire attitude might work toward errant infants, but we do not find much evidence of cats that are left alone conveniently herding themselves into a paper bag! Not all *nots* are consistent!

The *not* perspective is a case where one point of view held by a stakeholder is inverted and a contrary point of view adopted by a second stakeholder. This can occur in a football team or president's cabinet, but it does seem to represent a sad loss of franchise, on both sides.

For example, on the issue of withdrawal of U.S. troops from Iraq, it is argued that to post a timetable is simply to invite the enemy to hunker down, bide their time, and, when the United States has gone, continue their aggression at even higher levels of atrocity, presumably until they have achieved their goal. So what our enemies want is for us to withdraw. Accordingly, that is precisely what we should not do. Does this constitute a loss of vote? We do what our enemy does not want and do not do what our enemy does want us to do? So our enemy does not want us to stay in Iraq? If this is true, he is certainly gaining much from what he does not want: the ability to increasingly improvise IED warfare; the ability to propagandize on the success of that; the ability to demonstrate to a watching world that guerrilla-type warfare is incessant, relentless, and tough to beat; the ability to expose the relative impotence of the world's sole surviving superpower in the face of what is mildly termed an *asymmetric threat*. None of the foregoing constitutes a sufficient argument for withdrawing, but it does show a complexity in contrast to what an elementary rendition of the *not* perspective would belie.

But this perspective is not without merit if one handles it correctly. For example, it can be exceedingly useful in the case of assumptions review. Consider the following account:

> Greg and Tracy lie dead on the floor. Eyes wide open. Near the lifeless bodies are small puddles of water and pieces of broken glass. But there is not a trace of blood to be found. Further chards of glass lie on a table that sits beside an open window. A casement swings to and fro as the breeze blows, fluttering the curtain. How did the poor couple meet their death?

In order to come up with a reasonable explanation for the demise of Greg and Tracy, we have to be able to piece together the evidence that is available and our own case experience of crime reports that bear similarity to the situation described. Inevitably we will make several assumptions and proceed further on the basis of these at risk. The question we now pose is: What assumptions do you make as you navigate this scene? Make a list, as tedious and pedantic as this may seem. Then use the *not* perspective on one or more of these assumptions. We are not going to give you our answer, but we will give you a clue.[3]

So what are the thought foci for the *not* perspective? Well, the entities we have worked with so far can serve us again—candidates, need, and mechanism. In this case our candidates need to be identified in terms of making explicit what assumptions we are relying upon to meet our need, or perhaps to counteract the opponent's need or goal. We may even have to make explicit what we believe our enemies' assumptions are. The mechanism we need for the *not* perspective is tantamount to inversion, though it is not clear what that might yield. Some *nots* are not so obvious. What this does is to produce a variety of different needs and perceived needs of the enemy. The entities interact, as should all elements of a system, and produce a different approach to the initial set of need, goals, and assumptions. None of this takes away the need for intelligence, experience, and common sense, but it does perhaps give those things a better chance of resolving the problem, solving the puzzle, or unraveling the situation.

Chapter one: Perspectives

1.2.4 Paradox

We come to the final kind of perspective and by far the most intriguing. It is a case of *and* and *not*. Just to complicate matters, it is also a case of *not or*! This case makes us think more than any other. The risk is that the thinking will do us no good and so we reject it as pointless and unfruitful. The true test this perspective brings is whether we withdraw from it or find a way through it. Our development of foci for the previous three perspectives will hopefully find more useful service and so help us find a way through.

Here is something to think about. Consider a sheet of paper that has writing on both sides. On one side appears this statement: "The statement that appears on the other side of this sheet of paper is true." No ambiguity there. We turn over the piece of paper and find a second statement: "The statement that appears on the other side of this sheet of paper is false." Once again, a simple, clear statement. But wait. If that second statement is true, then the other statement, the first one, is false. Correct? In which case that first statement is saying that the second statement is false! But we just asserted it to be true.

Let us start again. Take the first statement to be true. If that is the case, then the second statement is true. But if the second statement is true, the first statement has to be false. But we just … and so we get lost once again. How can such ambiguity arise when there is none in either statement? Is it possible to get something from nothing? Put another way, can the whole be greater than the sum of the parts? There is a whole here and it is not just the sheet of paper. The whole is identified by the fact that these statements relate to one another, they refer to one another. So there are parts (each statement), there are relationships (what each part says about the other), and there is now a whole. And that whole is (an example of) the *paradox* perspective.

Paradox is the subject of an entire chapter, so our purpose here is necessarily confined to its essence and to its representation in terms of perspective. Moreover, we pose the question: How often do we find among multiple perspectives the potential for paradox? That being the case, do we reject the "offending" viewpoints because we do not want to face this unnecessary mystery of paradox, or do we esteem these in particular and look for wisdom from above, deeper insights into the situation that gave rise to these viewpoints? Accordingly, we represent two examples of a paradox and attempt to get some better handle on the meaning of this perspective.

Our first example typifies the impossible versus the imperative. Engineers may be familiar with this in that they often do the impossible dictated by their superiors or customers. On a more conceptual level, take the case of a flea leaping across a table by making jumps that bifurcate the distance yet to be covered. After one leap the flea is halfway. After two leaps there is only a quarter of the table to go. Three, four, five leaps and the flea is considerably nearer. The situation seems to be that he must surely get there but equally he needs to take an infinite number of leaps. From an imperative standpoint the flea must surely get there—there is nothing to stop him. But can an infinity of steps really be taken, in time? This paradox is solved by a brilliant piece of mathematics, the concept of the limit, which gave birth to calculus in the late seventeenth century, and for which all high school math students remain eternally grateful. But that calculus has not always been available to us. It took a leap of imagination into hitherto uncharted territory. It was wisdom from above, rather like the proverbial apple that allegedly fell on the head of Sir Isaac Newton, whose genius is credited for the invention of calculus.

A second example brings two very powerful forces into conflict, which so often happens, judgment and mercy. In civilized society we go to great lengths to legislate and to make effective judgment in accordance with law. When we abandon law we risk our very selves. But as our minds struggle with the formulation of law, including the appropriate penalties commensurate with the guilt of an offense, we have a deep inclination toward mercy, which is the deliberate setting aside of judgment. It is not a case of either-or, though we may make that choice; it is a case of both, and yet the paradox is you cannot have both. They rule against one another. It takes great wisdom to wrestle with both and come through the other side, as opposed to making a choice and possibly missing this access to wisdom. A superb illustration of this wrestling comes from a person famed for his wisdom, King Solomon. When confronted with a case in which two women laid claim to a newborn babe his decision was to divide the child in two and give half to each claimant. This judgment evoked mercy. While one woman remained silent, accepting the king's decision, the real mother cried for mercy and gave her precious child over to the other woman. The king knew that only the real mother would be willing to sacrifice her child to the false claimant rather than to the sword. Accordingly, he gave the child to this woman. Mercy triumphed over judgment, but only because the dilemma was confronted and wisdom afforded an entrance. Thank God for paradox, calculus, and wisdom.

1.3   *Perspectives on Google*

We both know a lot about Ph.D. students. Among the many we have been privileged to mentor some have been quite outstanding. In one case two students, who had been close friends from early childhood, after graduating went on to form a software company specializing in statistical process control for manufacturing enterprises. Today they are millionaires though their company's technical expertise does not relate to their Ph.D. work.

The Ph.D. process is not about getting another degree. And it is not about doing research. It is not even about developing leading-edge expertise to make your fortune, although few would pass up this opportunity. For us, the process is really all about the development of character. Over 3 to 6 years, depending on the circumstances, the intellectual challenges hurled at a Ph.D. student ensure a person's character gets developed. Resilience gets built in so you learn to take the knocks and come back swinging. Maybe even an agility gets embedded so that you learn to avoid a lot of the knocks by anticipating what is to come and making whatever reconfigurations are needed to swing around the blows and move on to higher ground.

We would like to turn our attention to two former Ph.D. students in the Computer Science Department at Stanford University: Sergey Brin and Larry Page. Both young men came from academic parents. Stanford is world renowned for its scholarship and entrepreneurship. One of us once asked a Stanford professor, "Can you get tenure on arrival here?" He blinked wondering whether it was a serious question and then reposted, "Yes. If you can walk on water." Some Stanford faculty may not be able to do this literally but in many fields they do it equivalently, and they are a mighty role model for their Ph.D. students. Page and Brin may not be walkers on water, but they were pretty good surfers, and in their own way they got out of the boat like Peter did.[4]

For some time the two had collaborated on search techniques to help users locate information on the growing World Wide Web. They published a paper outlining their algorithmic designs called Page Rank (it had started out life as Backrub) and had the benefit of their own university preferring this search engine for its own needs. The Office of Technology Licensing (OTL) at Stanford is no slouch. It exists to promote good ideas, borne of the outstanding thinking from students and faculty, in the marketplace in order to make money and allow faculty and students to concentrate on scholarship. The boys approached Stanford's OTL with their designs, which then promptly set about a brokering project to sell the IP, make some money for Stanford and the boys, and then let them get on with their Ph.D. programs. Alta Vista was offered the search technology for $1 million but turned it down. The boys were disappointed. Not so much by the loss of potential income but because the world was being denied an opportunity to benefit from their endeavors. So committed were they that they took the monumental decision to suspend their studies and leave the program to set up in business for themselves. Today the company built on that technology is worth over $200 billion. It is Google, Inc.

The phenomenon that is Google is nothing short of miraculous. From nothing to bigger than GM in less than 5 years. That gets people's attention. The Google story is today well told by many, and David Vise[5] has been our preferred narrator. Tales of Stanford venture capitalists dropping in $100,000 checks to the boys when they had no business model, and not even a bank account into which to pay the money, charm the senses. Spice is added when you consider that person's investment today is worth perhaps $500 million. We'd all take out a second mortgage to make a bet like that, but how would we recognize what Andy Bechtolsheim saw? We may be taking liberties here, but we want to suggest that some systems thinking went on in the development of Google, and with that as a lens, we explore more closely those elements that seem to us to be crucial to that success, and how perspectives played a part in the creation of those elements.

The first perspective concerns the structure of the Web. The prevailing notion was that it had no structure. How could it have? Anybody could join at any time in any way that worked for him or her. Standards and protocols applied for connection purposes, but as for central governance of the Web, its growth, the relative importance of Web sites, and so on, no structure was imposed and none could be said to exist. In such a case how could it be possible to be algorithmic about searching for information? The idea seems doomed at the outset. Along come Brin and Page. They believed there is a structure and their designs exploited this. That structure is similar to the citation of published work that is so familiar to academics and scholars. To show competence in a field you have to know what is being said, what papers are being published, and what the value of that work is. As part of any paper, references are made, intelligently in the body of the text, so that other people's work is cited. The good work is cited more often. So to demonstrate good work yourself, you are expected to cite the good work already out there. This, the boys believed, was the way the Web worked. Web sites cite other sites. The more the citations, the more important it can seem that cited site is. It becomes even more important if it is cited by the important sites. This became the basis of the search technology Page and Brin developed and remains that way today. This makes the Google site the one to use, the site to cite. And it comes down to being farsighted. It is about perspective, and the boys showed they had it more than most. They set out on their systems thinking course of assumptions busting and thought inverting, charting a way that millions follow today, and laying a foundation for a brilliant business model.

One of the premises that we believe Google assailed is that of ownership. Once again, the received wisdom was that this comes from a sale. For example, a person walks into a shop carrying a purse or wallet laden with credit cards and cash. She sees something she wants and buys it. You get her money, she gets the goods. You once owned the goods, but now you don't. She does, and you have her cash. That is how it works. That is how ownership is transferred. It is "How to Sell and Buy 101." Not for Google. It decided to sell nothing. Instead, it sought to serve for free. The customers pay Google nothing. Customers love that, and Google is rather happy too. It gives Google a warm feeling. It also makes the company a ton of money. How? That is the second example of Google's radical business model.

In the first example people left with their newly bought goods, some of the money no longer in their wallet and the rest still in there for the next shopkeeper. In the Google model, as people leave with no goods to speak of, they leave behind some of their money without even noticing it. It is the digital equivalent of the Artful Dodger,[6] who knew how to pick a pocket or two. Not all Google customers fall into this category. Only those who visit the discreetly located and subtly termed sponsored links located on the right-hand side of the search results page that Google provides in response to a user's queries. Every time someone clicks on one of these sponsored links

Chapter one: Perspectives

that Web site pays Google a few cents, let us say a nickel. It is estimated that around 15% of the users bother to visit a sponsored link. The vast majority of Google searchers are satisfied with what they get on the left-hand side of the search results. But if Google gets 100 million visits a day, that is 15 million customers leaving, for which sponsors are paying Google $750,000 per day. Neat business. Perspectives count. The right ones.

A third element of this fantastic business model concerns the notion, a traditional truth, that moneymaking is a one-shot deal. It interests us when we go to buy groceries (yes, men do that thing too) and stand at the deli counter figuring out what pasta, vegetables, and so on to buy. There is always a charming server on the other side of the counter who, having given you what you want, politely and invitingly asks: "What else?" Not having a shopping list, we always think, "I wonder what else." And so we come up with another request. We always get more than we need but never more than we planned since there never is a plan. It is an artful device. Google took this to a new level. The customers who leave their search results via the right-hand side, the sponsored link, are precious to Google. It is their interests Google is serving. Google does not simply settle for taking the sponsors' money, which is advertising money, and therefore there is risk for a sponsor in having captured the Google customer in its store; it still has to do a selling operation. The sponsor is in effect paying Google an introduction fee. But Google cares about its users and these customers. Passionately. So two factors govern the occupation of the sponsored links on the right-hand side of the results page. First, the sponsors have to bid, in a real-time auction. The successful bidders make it, but that success is predicated on a factor—the ability of that sponsor to be relevant to the customer, sufficiently so as to get him or her to unload money in traditional exchange for the goods. No matter how much a sponsor bids to get on the right side of Google, if they are not relevant, it will not work. Google does not take money from sponsors, no matter how much, if its precious users find the sponsors irrelevant. That is sharp, it is good business, and it keeps sponsors on their toes, making life easier for shoppers, and Google happy it is serving its users. If users do not leave money behind for Google, it could be because the sponsors squatting on prime real estate are up to no good. And they do not last.

The real estate remark is a fourth element in the Google systems think tank. Conventional wisdom is all about sweating your assets. Can you imagine a more valuable piece of virtual real estate in the entire Web than the Google home page? How much do you think Google can make from Coke or Budweiser for a 30-second banner? Are we talking Super Bowl time? And that event is only once a year. Google is 24/7. You do the math. Google keeps a clean sheet. It draws a blank from that real estate. It is the simplest site on the Web. Why? Because Google cares about its users. It exists to serve them, to help them find what they want, for free. If that is not the paradox perspective writ large, we do not know what is.

This philosophy points to a fifth element: if you want to get rich, you have to focus on money. What is more, rich dads make money work for them, poor dads work for money. Google works for people. Its users and those yet to become users are all Google cares about. The money takes care of itself. And so it has proved. Google misfired on Gmail. It provided a free service that looked great and typical, but then it peered into e-mails and tried to piggyback advertising having mined the semantics. Users balked at this and Google U-turned. The trusted servants risked losing a ton of trust chips and learned a big lesson. It proved the boys were not flawless. If they had gotten out of the boat and started to walk on water, when they lost their focus and sensed the enveloping storm they began to sink. They saved themselves by regaining focus. After all, they are only human. Well, superhuman.

There is one more element, one final perspective, on the Google business model worth noting, and that is to do with competition and cooperation. Google discovered the portmanteau word *co-opetition*[7] and played it beautifully. In this world there are neither competitors to fight nor cooperators to fawn to. Every enterprise in your business landscape is capable of both personalities so far as you are concerned. Your best bet is to understand their core competence and figure out how yours fits with that, in multiple shots. It is everlasting *quid pro quo*. So Google's tool bar ends up on the AOL Web site. AOL gets Google stock, cash, and a revenue-sharing agreement. Google gets the millions of AOL users hooked into its search engine.

Our parting remark on this retro analysis of a twenty-first-century corporate phenomenon using a systems perspective is this: people know a good thing when they see one. Really? Evidently Alta Vista did not. And maybe this foresight is the key to anticipating future events. What are the good things that people are failing to see and what is the significance behind this? Does it influence innovation strategies? Does it determine mergers and acquisitions? Does it affect Web site design? No answers. But maybe the right questions. That is what perspective is all about.

The boys did not find any answers to their Ph.D. theses, and still have not. But they asked the right question: Should we take what we know to a waiting world? The Ph.D. process is about building character, one strong enough to know that it is not about letters after your name. In the boys' case, it is about lots after your name. Think about it—*Google* is now an English verb.

1.4 Returning to Iraq

It is so easy to ask the question "Should we pull out of Iraq?" It makes the choice look simple, yet we know the decision is a tough one. For some it is straightforward: "We stay until we get the job done. We're still in Germany." People who favor the opposite decision can be equally economical in their response: "Let's get the heck out of there and leave them to their own devices." Polar opposites who would disagree that it is a tough decision. The undecided see the complexity. It lies in many more questions that cannot be shirked, that arise from that one simple choice of staying or leaving. If we pull out now aren't we leaving them with a mess of our making? After all, aren't they in a worse condition now than they were under Saddam Hussein? Don't we owe it to them to leave behind some sense of stability?

By contrast we might ask: But we surely can't stay forever? Can we really afford to be the world's policeman—world without end? We are the ones paying for this—if the world wants stability, they will have to pay their fair share. Will they? And if they won't, can we insulate ourselves from the inevitable pockets of instability that will form without this policing? Pockets that can still show up in our cities? Is there such insulation?

The debate seems endless. People tire. They make their choice, simple or not. Time's up. Move on.

And move on we will. The rest of the book awaits. But to conclude this opening chapter we ask two questions, and we answer them. First, what does it mean to be a systems thinker? And second, what does it mean for a systems thinker to move on? We need an answer to this first question because we want to encourage people to be systems thinkers. We think that systems thinkers stand a better chance at dealing with problems, especially the kind that this twenty-first century presents.

A systems thinker recognizes that a simple choice may hide a complex problem. We do not want to find complexity where it does not exist. Neither do we want to overlook it when it lurks in the dark. A systems thinker understands that complexity is made up of lots of things that interact with one another in curious, counterintuitive ways. This makes decomposition, the disaggregation of these various things into their separate existences, less useful. If we separate them we no longer have them, because they exist together. Our book will provide mechanisms for handling these things as individual things and as things that are together. A systems thinker understands the validity of individual perspectives, personal and subjective viewpoints on some objective reality.[8] A systems thinker need not choose from these but may find means to handle their simultaneity, notwithstanding their evident contrariness. A systems thinker respects the reality that perspectives can shift. This is more than someone simply changing his or her mind. A shift in perspective can be a powerful thing. One can sincerely ask why and find insight thereto. Looking at things differently can be beneficial, for oneself and for others who may be involved in your perceptions and by your perspectives. Mathematicians, scientists, logicians, theologians, artists, and philosophers have produced major breakthroughs in their work by leveraging slight shifts in perspective. It is not something that should be forced, but it is something that can be encouraged. We were hugely interested to discover that paradigm shift, a popular term much loved by management gurus seeking to encourage people to think differently, has the same meaning as repentance. It was a breakthrough for us both to realize that this word is not merely about being sorry or changing behavior, never an easy thing to do, but about thinking differently—having a mind change.[9] A systems thinker values perspectives and loves the fact that they change!

Finally, what does it mean to be a systems thinker to move on? To answer this we can find no better source of inspiration than that provided by C. West Churchman.[10] He argues compellingly that systems thinking is an endless cycle of perception and deception. We are mystified and so we search for truth. We do not know whether to pull out of Iraq or to stay. So we search for an answer. We find the answer. We were deceived all along by the mistaken question, or the prejudices, or the facts, or the news. Now we know. We have perceived. We determine and we live with the consequences. We move on. And then we realize that our perception was flawed. We were deceived yet again and we are once more mystified. And so we search. A systems thinker endures and enjoys this endless cycle of perception and deception. That is how he moves on. That is how she sees and sees that she does not see. Behind every perception lies a deception and beyond that lies greater perception. A systems thinker moves on perpetually and loves the journey. If you have learned anything from this first chapter we are thrilled and delighted. And so must you be. But you are equally deceived. Does this frustrate or disappoint? Don't let it. Be a systems thinker. Know that beyond your current deception lies fresh, exciting valuable perception.

1.5 Time to Think

1. Choose one of the topics below and apply the *or* perspective to develop a line of thinking for:
   a. A man contemplating marriage
   b. A woman seeking a new job

   Summarize your thinking in a two-thousand-word essay that meets the need of helping at least 20% of such people to think differently than they would have.

2. Consider the design of a jail. Different viewpoints exist as to the purpose that such an institution serves, including correctional facility, rehabilitation opportunity, incarceration stronghold, and punishment regime. Use the *and* perspective to develop a line of thinking that accommodates multiple viewpoints and yet realizes a single design. Summarize your thinking in an essay not exceeding three thousand words.

3. A pilgrim is making his way on life's journey to heaven. He reaches a fork in the road knowing that one way leads to heaven and the other to hell. At the fork live identical twins, one of whom always lies and the other who always tells the truth. The pilgrim knows this but cannot tell which is which. Both twins know which road leads to hell and which to heaven. The pilgrim can ask a single question in order, then, to be certain which road to take thereafter. What is it? What usefulness, if any, did the *or*, *and*, *not*, and *paradox* perspectives have in formulating a suitable question?

Chapter one: Perspectives

4. Let systems thinking be considered a strategic asset in Google and assume that much credit is accorded it by Google to its unparalleled success as a business. That being the case, how and about what must Google think, and what must Google do to avoid future pitfalls?
5. Suppose the forty-fourth president of the United States is a systems thinker of the first rank. Reproduce as best as you are able the inaugural speech of January 20, 2009, using exactly 1,776 words.
6. Greg and Tracy lie dead on the floor. Eyes wide open. Near the lifeless bodies are small puddles of water and pieces of broken glass. But there is not a trace of blood to be found. Further chars of glass lie on a table that sits beside an open window, a portal of which swings to and fro as the breeze blows, fluttering the curtain. How did the poor couple meet their death? (The point of this exercise has less to do with finding an answer and more to do with a careful examination of the line of thinking used to come up with an answer. We encourage you to think about not only making sense of the evidence, but also being creative about scenarios that might fit this evidence. We invite you to explicate each and every assumption you make in considering this scene.)
7. The Towers of Hanoi puzzle has many facets to it. The form of the puzzle is that there are three poles (or towers) and several discs of different diameter, each with the same sized hole in its center, a hole that enables the disc to be placed on any pole. The function of the puzzle is to transfer a neat pile of discs from one pole to another, one at a time, ensuring that at no time a disc is placed on top of a smaller one. The figuring of the puzzle is to create an elegant solution, one in which no mistakes are ever made and the discs transfer is achieved in the minimum number of moves. If possible, this elegance should be captured mathematically or algorithmically. Additionally, we have a further challenge to issue. We invite you to create an elegant solution space in which all of the various legal states of the puzzle are assembled into a framework such that the path of the minimum-number-of-moves solution strictly traces the boundary (an outer edge) of that space. You should be able to find an underlying pattern between this solution space and the algorithm that captures the solution. If anything, this exercise is a test of seeing what we do not see.
8. The paradoxical nature of freedom is illustrated by the following quotes. Use these and other examples from your experience of freedom in the world of work to develop a better comprehension of both freedom and paradox. Summarize your understanding in a three-thousand-word essay.

> Freedom is not doing what you want, freedom is wanting to do what you have to do … this kind of freedom is always rooted in practiced habit.

A liberal may be roughly defined as someone who, if he could stop all the deceivers from deceiving and all the oppressors from oppressing merely by snapping his fingers, wouldn't snap his fingers.

Men cannot escape from obedience to God. The only choice given to men, as intelligent and free creatures, is to desire obedience or not to desire it. If a man does not desire it he obeys, nevertheless, perpetually, in as much as he is a slave to his instincts and passions.

Endotes

1. See, for example, www.usip.org/isg/.
2. De Morgan's theorems meant that NAND logic formed the basic building block in all digital logic circuitry.
3. We both own pets. One of us has a dog and a cat, which showed up in the strangest circumstances on the property. Their names are Freddie and Sophie Boardman.
4. Matthew 14:30, Holy Bible.
5. Vise, D. A., *The Google Story: Inside the Hottest Business, Media, and Technology Success of Our Time*, Bantam Dell, New York, 2005.
6. A fascinating character from *Oliver Twist* by Charles Dickens.
7. Brandenburger, A. J., and B. J. Nalebuff, *Co-opetition*, Doubleday, New York, 1998.
8. "We can offer no greater challenge to pursue a deeper understanding of the relativities between subjective perspectives and objective reality than to recommend." Nagel, T., *The View from Nowhere*, Oxford University Press, Oxford, 1989.
9. The Greek word for *repentance* is *metanoia*—meaning a radical change of mind.
10. Dell, C. W., *The Systems Approach*, Bantam Dell, New York, 1984.

chapter two

Concepts

2.1 Just a Thought

We sit with our laptops, read e-mails, surf the Web, use office tools—word processors, spreadsheets, and presentations—*while* listening to our favorite pieces of music spinning from the E-drive. Words emerge from our fingertips, dance before our eyes, and fill our hearing. The Bee Gees sing: "It's only words and words are all I have to take your heart away." Our souls are stirred, in either melancholic reminiscence or unprecedented creativity. Maybe we should call it the word-wide web. Where would we be without words? Listen to a former prime minister of Great Britain:

> I have, myself, full confidence that if all do their duty, if nothing is neglected, and if the best arrangements are made, as they are being made, we shall prove ourselves once again able to defend our Island home, to ride out the storm of war, and to outlive the menace of tyranny, if necessary for years, if necessary alone. At any rate, that is what we are going to try to do. That is the resolve of His Majesty's Government—every man of them. That is the will of Parliament and the nation.
>
> The British Empire and the French Republic, linked together in their cause and in their need, will defend to the death their native soil, aiding each other like good comrades to the utmost of their strength. Even though large tracts of Europe and many old and famous States have fallen or may fall into the grip of the Gestapo and all the odious apparatus of Nazi rule, we shall not flag or fail.
>
> We shall go on to the end, we shall fight in France, we shall fight on the seas and oceans, we shall fight with growing confidence and growing strength in the air, we shall defend our Island, whatever the cost may be, we shall fight on the beaches, we shall fight on the landing grounds, we shall fight in the fields and in the streets, we shall fight in the hills; we shall never surrender, and even if, which I do not for a moment believe, this Island or a large part of it were subjugated and starving, then our Empire beyond the seas, armed and guarded by the British Fleet, would carry on the struggle, until, in God's good time, the New World, with all

its power and might, steps forth to the rescue and the liberation of the old.[1]

Yet more words. Just words? Not at all. Never! (as Winston Churchill himself said). Words are not just words, in some contexts. They are power and strength. They are purpose and will. They are fuel to the fire of our directed lives, and they are balm to the hurts of our dejected lives. They are the wings of our ideas or the wind beneath the wings of our thoughts. Here are more words, from a former British citizen:

> We hold these truths to be self-evident, that all men are created equal, that they are endowed by their Creator with certain unalienable Rights, that among these are Life, Liberty, and the pursuit of Happiness. That to secure these rights, Governments are instituted among Men, deriving their just powers from the consent of the governed.
>
> That whenever any Form of Government becomes destructive of these ends, it is the Right of the People to alter or to abolish it, and to institute new Government, having its foundation on such principles and organizing its powers in such form, as to them shall seem most likely to effect their Safety and Happiness. Prudence, indeed, will dictate that Governments long established should not be changed for light and transient causes; and accordingly all experience hath shown that mankind are more disposed to suffer, while evils are sufferable, than to right themselves by abolishing the forms to which they are accustomed. But when ...
>
> We, therefore, the Representatives of the United States of America, in General Congress, assembled, appealing to the Supreme Judge of the world for the rectitude of our intentions, do, in the name, and by authority of the good People of these Colonies, solemnly publish and declare, That these United Colonies are, and of Right ought to be Free and Independent States; that they are Absolved from all Allegiance to the British Crown, and that all political connection between them and the State of Great Britain, is and ought to be totally dissolved; and that as Free and Independent States, they have full power to levy War, conclude Peace, contract Alliances, establish Commerce, and to do all other Acts and Things which Independent States may of right do. And for the support of this Declaration, with a firm reliance on the Protection of Divine Providence, we mutually pledge to each other our Lives, our Fortunes and our sacred Honor.[2]

Words not only convey our intent and stir our hearts, they change our thinking, indeed our very lives. They define our journey and nourish us through it; they set a course and a destination; they can both arrest us and propel us. What are the words of the systems journey? How do they identify, locate, and uphold us? Inspire, inform, and instruct us? What are the tough words of correction and reproof? And the tender words of hope; to recover, reform, and rediscover the journey's joy?

In this chapter we give you our words, words that we believe are key to forming a foundation for thinking about systems—a thinking that helps us to understand these systems, whatever they look like and wherever they are found, that helps us to deal with them, by managing, (re-)designing, or disposing of them.

2.2 What's the Big Idea?
2.2.1 E Pluribus Unum

In the preface of his marvelous book *Complexity*, M. Mitchell Waldrop poses a variety of questions among which are these:[3]

- Why did the Soviet Union's forty-year hegemony over Eastern Europe collapse within a few months in 1989?
- Why did the stock market crash more than 500 points on a single Monday in October 1987?
- Why do ancient species and ecosystems often remain stable in the fossil record for millions of years—and then either die out or transform themselves into something new in a geological instant?
- Why do rural families in Bangladesh still produce an average of seven children apiece, even when birth control is made freely available—and even when the villagers seem perfectly well aware of how they're being hurt in the country's immense overpopulation and stagnant development?
- What is life anyway? Is it nothing more than a particularly complicated kind of carbon chemistry? Or is it something more subtle? What is a mind? How does a three-pound lump of ordinary matter, the brain, give rise to such ineffable qualities as feeling, thought, purpose and awareness?

These are all interesting questions and could occupy a thinker for quite a period of time. But this inspiring author asks a much more challenging question, a meta-question: What do all of these questions have in common? To read his book is to embark on a fascinating journey replete with insights and answers.

The reason we reproduce that theme, of commonality when faced with variety and the apparent lack of anything common, is because we believe this is fundamental in the phenomenon that is systems. Our line of questioning goes something like this: What is similarity or sameness? Does it exist or do we imagine it? When we discover it—even among obviously different objects (classes, behaviors, types, call them what you will)—what do we do with it, how do we leverage this sameness? Variety, they say, is the spice of life. What then is the purpose of sameness? Not the identicality of objects, for nothing is identical, since it is a slave to time, which is never the same. But in the notion of being somewhat similar and therefore familiar is similarity for our comfort, in a world of continuous variability? Is similarity the key to unlock our story, the story of ourselves and of the world we inhabit, both created and invented, a story that is driven by continuous change and endless variety? Is sameness *the* answer to the conundrum of many *and* one? And if so, isn't its meaning worth discovering?

2.2.2 The Sameness of Systems

Gerry Weinberg writes: "Every model is ultimately the expression of one thing we think we hope to understand in terms of another that we think we do understand."[4] Let us go with this, for the time being. *Thing* is a valid term, as are *think*, *hope*, and *understand*. But this sentence, which we find meaningful, cites the notion of model, as a device for comparison—of two things—for the purpose of extending understanding. Understanding may be a place, a state, a location, or some point, but extending understanding is a journey. Some of us have come to a stop, sadly. Others race ahead and maybe miss some important sights. Yet others are deliberately slow and so miss great events that expire before they get there. No matter the speed, provided it is of nonzero value, the game is afoot and we are on a journey, of mystery, discovery, disappointment, delight, and destiny. And model is a valid means of transport.

We are drawn to the idea that a system is one kind of model. In fact, a special type of model that equips it to be properly compared to certain things, for the purpose of extending understanding. We do not believe that all things are usefully likened to systems. Some believe that system is just another word for thing—but far less precise in meaning! In fact, we will now argue for a precision to the meaning of system that will make sense to compare some things to a system and not others. We base our arguments on an understanding of ideas, studies, and results that have come from many thinkers and practitioners.

For a model to be a vehicle for extending understanding it must have form, it must have function, and—even if it is wrong (whatever that means)—it must be useful. For these reasons, so must a system have form, function, and utility, at least for it to be a model. By form we mean shape or structure, and by function we mean behavior or dynamism. By utility we mean value. In *Pretty Woman*, the delectable Vivian (played by the rousing Julia Roberts) replies to a primitive question from her adopted escort Edward (played by the urbane Richard Gere) with another question. "What's your name?" he asks. "What do you want it to be?" she responds. It is context that tells us her response is not suggesting arbitrariness. It is simply suggestive. He is her client. She wants to delight him. Provide value. Be useful. She already has form and function. Now she integrates them. Model play.

There is something canonical about this trio of form, function, and utility. For that matter, trinity has proved a foundational term for many. One of us is the proud owner of a Mercedes Benz. The sight of that company's logo, enthroned on the hood, on long journeys is a continual source of intrigue. What does one see? An encircled star? Three discreet radial spokes with three arcs standing for the relationships between each one, blending into an infinite single circle. Three plus or minus zero produces seven. This would not be the case if we disallowed relationships, or if we could not perceive unity, and at the same time the parts that make up this wholeness. And so our fascination for threesomes, which strongly suggest integration, has led us to sum up our findings about *system as model* as a collection of proven trios, any one of which is capable of steering our thinking in useful directions.

2.3 The Conceptagon

2.3.1 Boundary, Interior, Exterior

There is something simply inescapable about the notion of boundary. Even when we cannot see one that in some way or other encloses space (in however many dimensions), we can still believe that there is something out there—beyond the boundary.

We accept infinity as a concept and as a useful artifact in mathematical calculations, but we can still be unhappy that this should represent a limit. There is always more. A motel that has an infinite number of rooms and is full can always make room for an infinite number of new guests—you just put all the existing guests into the odd numbered rooms and you suddenly have space for the extras!

The boundary separates the outside from the inside and we are glad to know that there is a difference; there is a peace in separation. Boundaries define the area of responsibility and the scope of interest. They tell us what are the things people can do something about (or not) and the things we can be properly focused upon (or not). Boundaries can be expanded or shrunk; in a sense they can be eliminated, making extinct what once was and lives on only as history. In that sense the boundary continues but changes its shape over time, depending on who is writing and reading history.

Boundaries can be convex, defining space in which interpolations between two interior points also belong to the interior; or they can be nonconvex and perplexing, where it is not only dangerous to extrapolate (it always is) but also risky to interpolate. Boundaries can be transgressed, calling attention to those who legislate for, police, or own the boundaries.

Boundary is an essential part of form, and system boundary is an essential construct for the system as model. With it we not only establish focus (system of interest [SoI], also known as system under observation [SuO]) but also begin to define context, be this wider system (or external system) and environment. These notions usually presume convexity, which leads to hierarchy that enables us to be more fluid in our thinking. However, reality punches us in the nose, leaving us, inter alia, with a mess of multiple inheritance conflicts. But then reality is under no obligation to conform to our model of it. Nor, for that matter, need the model conform to reality, though that is the conventional paradigm.

Not all transgressions are illegitimate. Indeed, some are essential. The human cell, whose boundary is sharply defined by wall and membrane, cannot be sustained unless there is substantive flow from exterior to interior and vice versa. The regulation of this flow is a sign of a healthy cell. So it is with humans. John Donne was being more than poetic when he wrote:

> All mankind is of one author, and is one volume; when one man dies, one chapter is not torn out of the book, but translated into a better language; and every chapter must be so translated.... As therefore the bell that rings to a sermon, calls not upon the preacher only, but upon the congregation to come: so this bell calls us all: but how much more me, who am brought so near the door by this sickness.... No man is an island, entire of itself ... any man's death diminishes me, because I am involved in mankind; and therefore never send to know for whom the bell tolls; it tolls for thee.[5]

At all levels of society, from individuals through groups and corporations to nation-states, the legitimate transfer of substance—atomic or digital—is not only desirable but needful. The challenge to systems people is to insist on a boundary while simultaneously insisting that it not be.[6] Boundaries, in the systems engineering organizational sense, are essential to the maintenance of expertise. But the expertise exists for a purpose, which is that it be rendered as a service to others, from whom learning can be obtained, explicitly or otherwise. Solid boundaries rigidify expertise, turning the interior into a stove pipe and causing ill-health in the exterior parts. Never has there been a greater need for intelligent walls.

2.3.2 Wholes, Parts, Relationships

A blob is not a system.[7] An amorphous viscid lump of material has precious little form, no obvious function, and zero utility—at least at face value. However, if the blob is slime mold, though this might appear to be a formless, lifeless lump, it would certainly be considered a system. Slime mold cells, though relatively simple, have attracted a disproportionate amount of attention from embryologists, mathematicians, and computer scientists because they display such intriguingly coordinated behavior, about which we will say more below. The point at issue right now is that this blob is not an amorphous mass but rather a collection of parts (cells) with relationships between them (chemical exchanges) formed into a whole (slime mold blob)

Chapter two: Concepts 25

with a behavior that is inexplicable at the level of the parts. The whole in some sense can be compartmentalized, except that it cannot be explained using reductionism.[8]

We have observed an extended use of the term *systemic* in recent days, mainly in the attribution of failure.[9] Thus, while a part of a system might fail, for example, a signal passed at danger by the driver of a passenger train, this in and of itself does not make for a catastrophe, though it is an incontestable contribution. It is the interconnection of several parts of the system that constitutes the overall failure. In the case of a rail disaster leading to multiple fatalities, it is the lack of an automatic braking system, the high traffic density, the absence of safety features in the event of an accident, and poor communications in alerting emergency services to attend the scene of an accident in rapid response.

We perceive two imminent challenges for systems people when we look at this particular triad. First, the tension established on behalf of a part (of a whole) to essentially belong to a whole, for that is its reason for being, while simultaneously being the part that it is (or the whole that it is), for that is why it was chosen or formed in the first place. The tension is between belonging and being autonomous, of being independent and yet interdependent, not codependent. It is a real and not imagined tension, for diversity is what gives the whole its strength and its function, and yet harmony is what the whole needs for its form.

The second challenge is related to the first and has to do with the term *system of systems* (SoS). In this setting, the SoS would be considered the whole and the systems its parts. Conventionally, these parts never intended to be so. No one imagined that they would become parts of a greater whole; if that were the case, this emphasis on interoperability would not be so great or its challenge so huge. Now the tension is between not belonging (in the first instance) and making changes to the parts (the original systems) in order to be able to belong.

The systems movement has historically emphasized wholeness. With these challenges in view, the pendulum is swinging toward partness.

2.3.3   Inputs, Outputs, Transformations

Scientists assert that energy can be neither created nor destroyed, only changed in form. In other words you cannot get something for nothing, but you can get something different, for a price. Engineering is the profession that seeks to increase the variety of transformations while decreasing the price to pay. With this as a context, it is inconceivable to eliminate or even to overlook, as concepts, these transformations and their nature, and the things being transformed, the inputs and the outputs. They are in our world.

One of us started out academic life as a control engineer, someone whose job it is to formalize this triad of concepts in ways that cause the outputs to behave exactly as desired, regardless of the variability in the inputs and the transforming system. Building models of such systems is a necessary condition for developing control schemes, yet more systems that take the outputs of the transforming system, affecting transformations on these in order to produce a new set of outputs that then become controlling inputs on the transforming system. We will say more about these schemes in the final triad.

Of course these inputs and outputs transgress the boundary of the system, which becomes by interpretation the transforming means or mechanism. An interesting corollary with this triad is that these mechanisms or systems can be concatenated whereby the outputs of one system become the inputs of its neighbor, and so a series or sequence of transformations is possible. The control scheme referred to above is a special type of system interconnectivity with feedback being the differentiating mechanism. However, now we can conceptually open up a world of linked triads in which feedback loops are far lengthier than so far imagined.

One way, a type of medium, for capturing this world is the flowchart where each element of the triad is explicitly articulated, as shown in Figure 2.1. One thing to be noticed in this figure is that the same artifact is both an output and an input, and this has real added value. For example, it can now be seen to act as an interface between systems. In another way, it can be seen to justify the dependency of one transformation upon another, in the sense that the diagram is depicting process that with the correct information built into the elements of the flowchart can be automatically translated into a project plan, as depicted in Figure 2.2.[10]

The concatenation of systems in this way can be used not only to portray a sequence of tasks, but also to describe the discrete steps or distinct phases of an extended trip, for example, the systems journey. Howsoever these elements are pieced together, from the bottom up—reflecting experiences of journeymen—or from the top down, using the method of functional decomposition and step-wise refinement, the result is a systems map showing either a thought process or a required physical effort for making an orderly series of transformations of an initial input, say, an identified opportunity, through to a final output, say, a Web service to arrange honeymoons for desert beach lovers.

Some might call this an end-to-end business process (see example in Figure 2.3), others a method for turning a need or requirement into customer satisfaction, and yet others a value chain integrating the various specialisms needed to process a desire and make it a reality. The language is different but the sentiments are identical—separate into manageable parts and integrate into a whole, making absolutely sure that things join up, seamlessly if possible, for which outputs and inputs are key.

2.3.4 Structure, Function, Process

Let us take the last triad and apply it to the human heart, the physical object rather than the seat of affection about which poets are apt to write so ably.

Chapter two: Concepts

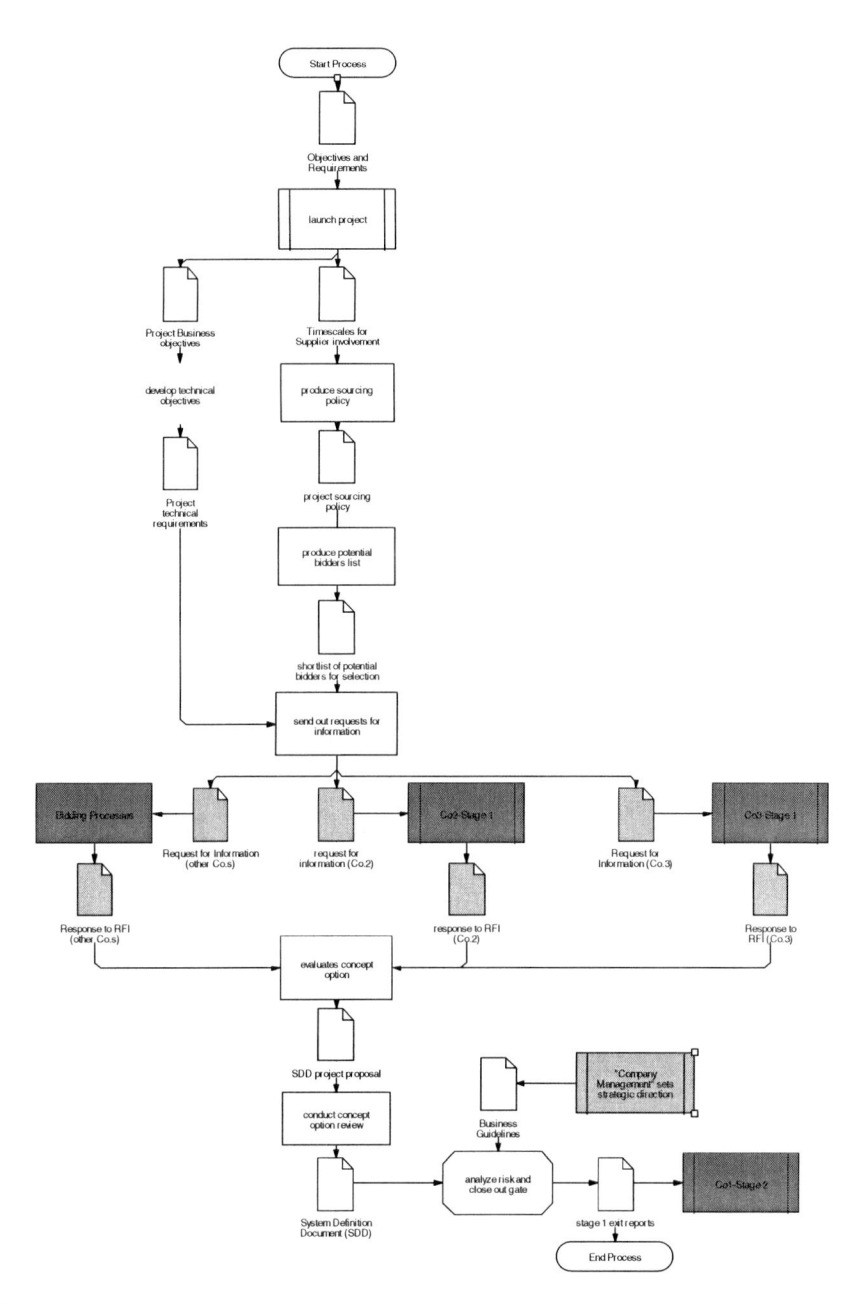

Figure 2.1   Flowchart of a project process.

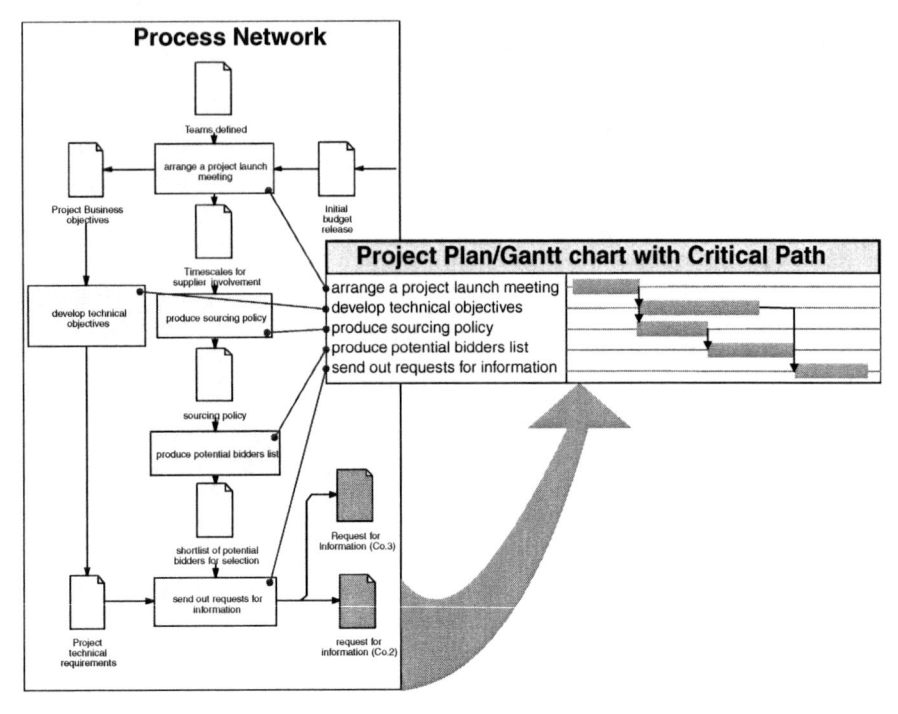

Figure 2.2 Process to plan. (Hammer & Champy, Reengineering the Corporation: A Manifesto for Business Revolution, 1994. Reprinted with permission of Harper Business.)

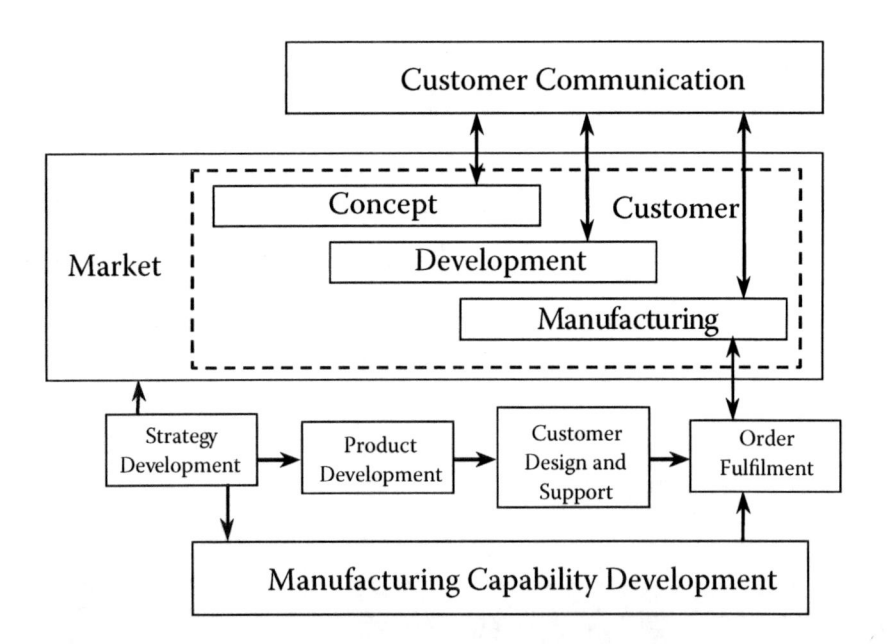

Figure 2.3 Texas Instrument semiconductor business process map.

Blood flows in and out. What transformation then? Evidently none. Except it is not just blood that flows, but blood without and with oxygen. The transformation takes place in another system, the lungs, but not without this system, the heart, operating as a pump. No pump, no flow; no flow, no oxygenation; no oxygenation, no life; no life, no nothing. Pumps matter.

We might say then that the function of the heart is to pump blood. If this shuts down, ceases to be, then worse will follow. It is not just a case of missing outputs, from the heart; it is a case of failed function. In order for this not to fail, the heart must continue to have a healthy structure. (Of course, this may be healthy and the function still fails for other reasons, but for now let us concentrate on the structure that enables the function.) This structure consists of four muscular chambers, various valves, and pipes that lead into (veins) and out of (arteries) the heart. This structure operates according to some processes that support blood flow: push and pull—alternate contraction and expansion of the chambers and co-coordinated opening and closing of valves pushes blood through the arteries and pulls it back through the veins.

Very quickly we move beyond the boundary of the heart, as a pump, into the wider system to which the pipes connect, soon recognizing that we are now dealing with a circulation system, of which the pump is merely a part, whose purpose is to provide an infrastructure for the exchange of matter and energy. This leads to a higher-level function, higher-level structures—to do with lungs, liver, kidneys, and so on—and higher-level processes, for example, oxygenation, dialysis, and so on.

This thought movement, guided by blood flow initially, takes us across an extended enterprise of bodily functions, structures, and processes, and it takes in a succession of system(s) of systems as we journey up the body's scale. This journey is facilitated by the notion of purpose or context, but it is activated by the interaction of a valuable triad: structure, function, and process[11]—one that attends a given system to which our previous triads may give their own independent testimony.

Structure defines components and their relationships; structures may possess both rigidity *and* flexibility, the one not necessarily being a contradiction of the other. Function defines the outcome (or desired outcome) or behaviors of the system. Process explicitly defines the sequence of activities and the know-how required to produce the function given the structure. Thus, these three form an interdependent set of variables that can define the whole, given a knowledge of that whole's purpose or context.

We have long since known that several different structures can realize the same function. For example, the automobile, the locomotive, and the aircraft, though quite different structures, each realize the same function of transportation. However, it is process that explains how a single structure can give rise to multiple functions. For example, the penitentiary, as a single structure, can give rise to functions of containment (protecting law-abiding citizens from dangerous criminals), correction (persuading inmates to become law-abiding citizens), and creativity (teaching inmates new skills and doing useful work during their period of confinement).

Some might say that the processes are in effect the functions of the components of the system. This would not be quite correct. The processes indeed comprise these component functions, denoting a change in scale—from system level to component level—but the processes tie these component functions together in a very real structural sense, although this is integration of behavior of elements rather than of elements themselves. What we do find, however, is the continual interaction of the forces of separation and integration, and the interdependencies of structure, function, and process across scales. This is the phenomenon of systems thinking. Greater depths of this thinking require attention be paid to these three: scale, moving across scale, and discovering new behaviors as we go to higher scales.

2.3.5 Emergence, Hierarchy, Openness

With the definition of system boundary we gained an interior and an exterior. Alongside the notion of interior there exists a sense that it can be governed. After all, what is the point of it being included if it cannot be reckoned on to help serve the purpose of the system defined by the boundary? What is inside the boundary is for the system and in common parlance needs to "get with the program." What about the stuff in the exterior, the outside content, the something that is out there? Does this serve the purpose of the system? Can this be governed? Is it with the program? Or is it opposed to what the system is trying to do or be?

This is one time we have to respond, "It depends." Sometimes the system boundary expands to embrace what formerly lay outside and now becomes part of the system. If it were that bad, why include it? Maybe it was only "bad" while it lay on the outside. When it comes inside it can be put to good use. Lyndon B. Johnson once said of Edgar Hoover, director of the FBI, "It's probably better to have him inside the tent peeing out, than outside the tent peeing in."[12]

Then again there are things that lie outside the boundary that are clearly advantageous to the system, a good enough reason for acquiring them. Sadly, the power that these additions were foreseen to bring does not always materialize and the system is poorer for the acquisition. So it goes. Whether for ill or good, what is exterior to the system cannot be ignored by the system, which is why the system is open to exchanges with the constituents of the exterior. This openness is an inevitable part of a system's being and behavior. To be closed to the exterior is to face death—and that right quick. To be open may result in death, but therein lies the way of life.

Closed boundaries are simply not an option for any system. While being too open is risky, a system can only learn what this means by being willing to be open in the first place, and then adapting its behavior toward future openness based on its experience with its formative exchanges with the exterior.

While adaptation is an important attribute for a system to have, whether these are organismic or not, so that it becomes more capable of enduring hostile exteriors, it is not uncommon for systems to search out friendlier neighborhoods where the exterior is less hostile and more conducive to the systems' development. What are these friendlier neighborhoods? Well if we were to call the exterior, at large as it were, the environment, then the notion of wider system can be conjured to suggest the friendlier neighborhood. One is the great unknown wherein may lie predators and agencies decidedly hostile to the system's well-being; the other, a more local and familiar region, is somewhere where the system has a good opportunity to grow and develop. And this development takes place in the vicinity of neighboring systems, all of which have "decided" that the wider system is the place to be, the place to belong.

There is not much that a system can do about its environment. Maybe in some case it can "lobby" along with others and so make the environment better, for it and the other lobbyists. But then there will be other systems in the environment lobbying for a different environment. In the end what becomes of the environment is a great unknown. However, this is not true of the wider system. In this case there is strong receptivity by it to the combined and concerted efforts of the like-minded systems belonging to it and indeed constituting it. In some cases the wider system is purposefully defined and the systems that belong to it deliberately placed there, by design. In other cases these wider systems actually form by the movement of the constituent systems tired of adapting to pernicious environments and more rewarded by relocating to friendlier neighborhoods.

Again this terminology applies not only to organic systems, for example, slime mold and ants, but also to intelligent systems, with nothing more than electronics and software to resource their adaptive skills. Openness becomes a powerful notion that systems of all types can exploit in order to do better, be better, and find better places to live. Openness becomes a portal for exercising other systems concepts such as boundary, wholes, exchanges (inputs and outputs), and process. It opens up new worlds, and closes down a few too.

When systems congregate into a wider system there is a need for organization, whether this be applied by prescient design (top-down) or by some self-organizing principles built within the constituent systems (bottom-up). The need for organization comes from the need for some stability of the wider system being formed, a stability that systems have, to some degree or other, within themselves in order to remain systems. A particular form of organization common among systems is that of hierarchy. That word today is strongly suggestive of command and control (which we will turn to later), of top-down planning, of superiority and servitude, and of feudal organizations in which the top is better and better not be at the bottom. This is our opportunity to straighten out hierarchies.

Of course those ideas are prevalent and not without precedent, but they have lost connection with the original ideas that make hierarchy so powerful.

We restate these fundamentals here and now, all of them to do with the dual of separation and integration. They are span of care, authentic relationships, management of healthy conflict, distribution of talent, and balancing perception with deception. When we consider these, we find hierarchies are less to do with organizational charts and more to do with vitality, survivability, and purpose. Let us do that.

Span of care means that no one is capable of looking after everyone else. Each has an ability to look after a few others, probably not more than seven—though the span factor is arguable. And everyone gets looked after. If the span is too great, the carer suffers and the cared for feel it. Disease spreads. Other carers, in good health, need to sense and respond to this premature demise of one of their own. And the cared for, under the span of the responding carers, need to respond also. Responsiveness spreads. And healing is at work.

None of this can work if there is not a regime of sensing and responding among the community of systems—the wider system of systems. Sensing and responding is all about relationships, a systems concept we already touched upon. But relationships need to be honest. For the conscienceless systems of hardware and software this should not be a problem—unless of course the designers of these have been dishonest, intentionally or mistakenly. But for people-centric systems, the opportunity to deceive, trip up, compete, and generally lack integrity is ubiquitous. Individuals may think they get away with being less than honest. Initially this may be true. But the community feels it, sooner or later. And the dishonesty rebounds on them, eventually. For open systems, whatever happens, the good, the bad, and the ugly, spreads like ripples or tidal waves around the wider system. And all are affected at some point.

Relationships can be tetchy. There again, for hardware and software this is less of a problem since clean interfacing removes ambiguity and tetchiness. It is as if all of this source of aggravation is transferred to the design of these interfaces and to the creation of standards that govern them! But for the belly-button systems, conflict arises naturally, from where, who knows? The response of some people is to nip it in the bud. Others let it run and run. Some use force to eradicate it, not counting the cost in the loss of humankind and kindness. Others ignore it, preferring "mercy" so that the genuine source of conflict is never treated. But conflict is inevitable. Healthy conflict is actually good for community. It is just that this is hard to recognize. The management of healthy conflict is a duty of a wider system, on behalf of its constituent systems, and the practice of managing healthy conflict brings health to the community.

Not only can no one look after everyone else, but no one can do everything. Each has his own gifting, his own talent, his own purpose. That is a ruling principle of community and an axiom in the design of systems, be it the belly-button kind or the regular push-button variety. Discovering this gifting in people and their groupings, or designing this talent into functional and physical architectures, is what makes for good hierarchies. Some

Chapter two: Concepts

systems are good at knowing what the gifting or talent distribution is—in other words, knowing where to find functional contributions among the spread of talent. Given this information they might then share it with those who have a gifting for getting these talented systems to render service as effectively, timely, and resourcefully as possible. Both these system types are less about making knowledge contributions to wider system behavior and more about knowledge that leads to the release of efficacious behavior. Never despise the contribution of "knowing someone who does know." Degrading gracefully at the boundary does not signal weakness but is rather a sign of health.

Balancing perception with deception is a notion that has never been better expounded than by C. West Churchman,[13] who characterized the nature of inquiry as an endless cycle of perception and deception. In rehearsing this we will not do justice; nevertheless, what it tells us is that in our current state of confusion we are able to perceive certain patterns and truths that lead us toward light and comfort and where we are able to remain peacefully and profitably for a while until we discover that what we saw was not correct. It turns out to be false. Our perceptions are in some sense flawed, deficient, or neglectful of new phenomena that are not adequately explained, and we are once again trapped in confusion, but perhaps a better state than when previous deceptive conditions held us. We inquire on and discover fresh perceptions and eventually with these new patterns and truths find fresh comfort. This entire cycle is humbling, a quality that serves the community well. It is character developing also because the search is perpetual, the comfort zones gratifying, but not so seductive that we feel a sense of arrival—only discovery. To the cynic, the cycle is nihilistic; to the immature, it is frustrating; but for the thoughtful, it is part of the systems journey.

Openness therefore leads to hierarchies, and hierarchies lead to emergence, and emergence is what makes the subject matter of systems different from the nature of science. Science became "king" when it finally overthrew the tyranny of religious ignorance, ironically governed by a hierarchy defined by the rule of priests. From future predictions based on spilling the entrails of goats, through dogmatic insistence on an earth-centric view of the solar system, to insufferable treatment of the uneducated and the disenfranchised, the power of churches over people has been in continual erosion. Science gained that power, ostensibly for all people, when it insisted on a value-free method for observation and a rational approach to interpretation and application.

Scientific method is based on repeatability, refutation, and reductionism. Peer review has been essential to good science, and as long as peers practice community as we argued for above, review works well. Hypotheses survive for as long as they resist refutation, by logic or observation. But the direction of science is always along the lines of reducing complexity to its constituent parts and insisting that phenomena are explicable in terms of the behaviors of simpler parts until those parts become irreducible. But this approach overlooks relationships and the very fact that when parts belong to a more complex whole, that belonging changes the nature of the parts from what would exist if there were separation versus integration. It is this very belonging that explains why complex wholes are more than the sum of the parts, and why phenomena are classed as emergent.

There is value in having systems of inquiry to serve different scales of complexity, hence the emergence of chemistry, biology, and psychology—each being sciences in their own right that require different bodies of knowledge to treat the phenomena that emerge at different scales. Physicists may insist that their science can transcend scales, from macro- through micro- to nanoscales. But that still leaves the existence of other sciences that attach to different levels of phenomena, finding explanation within their own bodies of knowledge. Some explanations of behavior, for example, at the macro level, can be traced through to the existence of properties and relationships at, say, the nano level—but not without crossing scales and navigating hierarchies, and this is now a systems movement.

What is indisputable is the reality that some attributes are meaningful only at certain scales and entirely meaningless at lower scales. Water has the property of wetness. But how can we say that hydrogen, a constituent of water, is wet? It makes no sense. Hydrogen is present in water, but is also clearly absent, while wetness only comes into existence with the water.

Some emergent attributes exist at multiple levels. For example, a person is smart; likewise, a group of people are smart. Even the eye of an individual has intelligence—or a form of processing power that gives the eye smartness. So intelligence exists in society, in persons, and in parts of the human body. But how do we figure the aggregation of intelligence from the parts? Some have argued that crowds show more wisdom than the wisest in the congregation.[14] Maybe we as individuals are smarter than we think? Maybe our consciousness lets us down at times? Maybe differential equations are only a partial answer for dynamics?

From openness to hierarchy and thence to emergence. Here is one final thought: Where does emergence lead? Some say it leads to chaos, which is, of course, abhorrent to the ardent admirers of order who insist on predictability even if this has to be through the lens of probability, where at least the outcomes are known and conform to a definite distribution. When this probability rule is itself subject to uncertainty, but can be governed by yet another probability distribution, that is termed by some as adaptive behavior.

A different way to look at this is to think of the outcomes as fundamental objects that exist in one of a number of simple states. These basic objects have rules built into them that are known and fixed, can refer to the condition of fellow objects, and determine the future states of the objects. The behavior of these objects, alternatively the future of outcomes, is now a product of extensive interacting behavior among the family of objects, and this results in patterns of behavior at scales above those of the objects, for example, within the hierarchy of objects, and the wider system as a whole.[15] These patterns are emergent, and we shift our focus to these patterns and away from the component objects. The emergent behavior could well be chaotic in appearance, possibly settling down to a stable state over a very long time, possibly not. The nature of chaos is aperiodicity. And yet we know that what gives rise to this chaos is order—the basic object or outcomes and the known, simple, and fixed rules governing future object states. In other words, we know there is a rational and simple explanation for complex and chaotic behavior. We have chaos and order—what some term chaords.[16] The systems challenge is to discover the simplicity in complexity, not by reductionism but out of respect for the myriad interconnections of the system's parts. The next step, which could in principle be any step, of the systems journey could well take us over the edge, from order to chaos. And it may not be possible to step back. Future steps may not make sense, they may well be counterproductive, and need to be counterintuitive. But that is part of the fun. And there could very well be, beyond chaos, more of the journey.

2.3.6 Variety, Parsimony, Harmony

"Samos is a magical island. The air is full of sea and trees and music. Other Greek Islands will do as a setting for *The Tempest*, but for me this is Prospero's island, the shore where the scholar turned magician. Perhaps Pythagoras was a kind of magician to his followers, because he taught them that nature is commanded by numbers. There is a harmony in nature, he said, a unity in her variety, and it has a language: numbers are the language of nature."[17]

This is how Jacob Bronowski, in his simply brilliant work *The Ascent of Man*, opens his chapter "The Music of the Spheres," his essay on mathematics as man's tool for discovering divine verities. If the systems journey must account for the conundrum of the many *and* the one, then harmony, that which achieves unity in variety, must be a fundamental systems concept. And we could hardly do better than allow Dr. Bronowski to facilitate its introduction.

Pythagoras identified the coincidence of numbers and harmony in sound, using a vibrating string for his experiment to determine the relationship of notes to nodal lengths. This notion of harmony, the governance of nature by numbers, was eagerly extended to the heavens in which celestial movements, controlled by numbers, had to be the music of the spheres. It was Johannes Kepler who resisted the temptation that planets moved uniformly in circles, the most perfect geometrical shape, and ventured to propose instead that these trajectories were elliptical and planetary motion nonuniform. That development spurred the mathematics of instantaneous motion in which time itself had to be subject to mathematical lucidity, for which breakthrough the notion of the infinitesimal step had to be formulated. With this the laws of nature became the laws of motion expressed in the calculus of Gottfried Leibniz and the fluxions of Isaac Newton.

But even as mathematics marches toward nonlinearity and chaos the notion of harmony is preserved and protected, because it is the only means by which variety and unity can be understood, how the forces of separation and integration are held in tension, and how complexity—regardless of degree—can be reconciled with simplicity. After all, harmony, we believe, is at nature's heart, this being for creationists a reflection of what its Maker desires and exhibits.

Engineers have always been smart enough to go with the flow. Though faced oftentimes with what scientists have asserted, with a cocksure arrogance, was impossible, the engineering profession has ridden the forces of nature—gravity, electromagnetism, aerodynamics, and others—to achieve their purposes on humankind's behalf. Fighting the force, going against the flow, spitting in the wind make no sense. Engineers have known the direction of the wind, not being able to explain "whence it cometh nor whither it goeth,"[18] and so accomplished their goal. Harmony is the key to reckon with natural forces. Clayton Christenson suggests that the same is true for societal forces.[19] How then can systems engineers leverage harmony in their push for innovative technological designs, including the human element?

Two systems concepts lie at the disposal of the architect to reflect the beauty of harmony: parsimony and variety. The law of parsimony states that given several explanations of a specific phenomenon, the simplest is probably the best. (Tina Turner would rejoice!) William of Ockham, a fourteenth-century English philosopher and Franciscan friar, is attributed with formulating this law, known as Ockham's razor: the simplest explanation to any problem is the best explanation. For the architect this boils down to the maxim "entities need not be needlessly multiplied." The engineer's equivalent is KISS.[20]

On the other hand, the law of requisite variety states that for a system to survive in its environment the variety of choice that the system is able to make must equal or exceed the variety of influences that the environment can impose on the system. Likewise, in order for a regulator to control the behavior of a system, for example, a hydraulics controller to operate the aileron on an aircraft wing, the controller must have at least as many degrees of freedom as the system it is seeking to regulate. Formulation of this law is often attributed to Ross Ashby and occasionally to John Von Neumann. Metrication of choice or degrees of freedom borrows extensively from the domains of information theory, games theory, and cybernetics.

And so the engineer who seeks to leverage or reflect harmony is hemmed in, for his own good, between keeping schemes as simple as possible and no simpler. Alternatively, to make systems absolutely no more complex than they need to be. Of course these laws do not tell us how they are to be observed, not unlike the Ten Commandments. As with all laws, transgression is a great tutor.

2.3.7 Command, Control, Communications

The triads we have looked at thus far have largely been observational. So we have a boundary, relationships, and exchanges with the outside (or other systems).

Chapter two: Concepts 37

So we have new forms of behavior, and criteria to observe in ways to design. So what? Making observations only takes us so far. They are like conclusions. But conclusions are not decisions. You can change your mind all you want. But you cannot change what you did. It is done. You have to do something else to undo it, if you can. And that is another thing you cannot change!

It would be wholly unsatisfactory to have no concepts that get us into the action, that make things happen, that make mistakes, from which we learn. Our final triad takes us there—into the world of systems in action—systems that talk to one another, and hopefully listen and respond; systems that control one another, and are suitably obedient; systems that give orders, that set directions and go there, or not. Without this last triad we would merely be bystanders, understandably fascinated and suitably informed. But with it, we are ready to roll.

The scientists have come up with the big bang theory to explain the origin of the universe. These guys are really under the gun. They are forced to figure out how something came out of nothing when one of their rules is that energy can be neither created nor destroyed—only changed in form. Suppose we make it easy for them. Suppose there never was nothing (sounds like bad grammar). Suppose zero has never existed, that there always was something, for ever and ever. Zero was invented to support the place value system in arithmetic. Suppose then that zero is an invention, a toy for us to play with, and nothing more than that. But there never was and never has been just nothing. Something has always existed, suppose. Then whatever that something was it certainly did not decide to become nothing. Instead it grew, and that growth either has come about from communications or has led to communications. The certainty of nonzero existence is communication.

Consciousness itself is a form of communication, with self, about self, and for self. But from this baseline has grown full-blown communications, a self-expression for the digestion of others to develop self, and sometimes so that others will benefit. This latter point may be altruistic or it may be disguised self-preservation since self benefits from being in a better environment made up of developed others. Whatever the motivation for communications, their existence seems to be as perpetual and persistent as nonzero existence itself. Losing the ability to communicate is a death knell.

One thing we have found, as practicing systems engineers, is the intricacy of communications, notwithstanding the law of parsimony. In order to pass on information, the heart of communications, inordinate preparations are required, dealing with pre- and postmessaging. Think about the basic communication elements: transmitter, receiver, channel, and content. Is this an exhaustive list? Certainly much more could be said about the channel in terms of bandwidth, security, and reliability—the classical terms of systems engineering. Likewise, content has dimensions for both syntax and semantics, and the latter always throws up issues of ambiguity and cultural alignment. Protocol governs more than the channel. Would you know how to greet the Queen of England?

What is missing from this list of four elements is acknowledgment. If I am ready to transmit, is the receiver ready? How will I know? Did the recipient get my message? How do I know? If I do not have the answers ready for this line of questioning, my attempts to communicate are likely to be nugatory. We have found it valuable to frame an acknowledgment template, whether we are transmitters or receivers, to satisfy ourselves that communication is working. This device serves much more than simply verifying that messages are passing to and fro. That is merely a baseline, and a source of huge misinformation.

Let us take a simple example. Suppose you need to know whether a particular package had been dispatched. How would you discover this? What communications would you enter into? You might go to the person whom you had told to make the dispatch and ask, "Did the package get dispatched?" She might reply, "I told Smith to send it." Are you satisfied? Did it get sent?

Perhaps you say, "I am not satisfied with that response. Do you know for certain that Smith did what you told him to do? How did you make sure?" Maybe if you are still unhappy you make this lady do her job to your complete satisfaction. Or maybe you go to Smith, making careful notes to do something about the gal's delinquency later in the saga, and have a conversation with him. Smith says, "Yes, sir, I took it myself to the dispatch department. Here is the receipt [acknowledgment template]." Are you happy? Maybe not. Maybe you ask, "Did you see the package loaded on the van? Did you watch the van leave the car park and head west on I-4?" What do you say? What are the communications that will end in your satisfaction that the package got dispatched?

For us, the acknowledgment template is one means for closing the loop on communications, which is by its very nature bilateral and maybe multilateral, and loops that are not closed reduce communication to broadcasting. At some point you trade off the design of an acknowledgment template against the value of communications, but you will not do that if it is missing from the communications structure. Systems thinking tries to find out what is not there, not just what is missing.

Loops are an indispensable commodity for systems thinkers. In a subsequent chapter we will examine the outstanding contribution that Peter Senge has made to systems thinking, beautifully illustrated by his use of influence diagrams to capture organizational dynamics with their myriad loops of causal relationships.

It is the profession of control engineering, however, that has done as much as any for making loops part of the systems thinker's toolbox.[21] The notion of feedback to regulate servomechanisms is the control engineer's contribution to understanding how systems can be sensed, and then sufficient sense made of this for the purpose of having the system behave agreeably. The cleverness of control has been to influence systems behavior when a priori knowledge of that system is difficult or impossible to achieve. Usually you need to know what it is you are controlling to have a chance of regulating its behavior; that is one consequence of the law of requisite variety.

Predictive control has been an outstanding contribution to knowledge and one of us was privileged to witness its birth. Brian Swanick and his Ph.D. student David Sandoz hit on the brilliant concept of using the control signal as the means of identifying the system by correlating this with the system outputs. Thus, observability and controllability coincided with the system being simultaneously identified and regulated. The systems were presumed to be multivariable, linear, and time invariant. However, even though these assumptions were not met in practice, provided that the perturbations to the systems were not massive, a linearized model, about some datum operating point, was adequate to control the system's behavior quite reasonably. Beyond this, more sophisticated schemes of an adaptive nature were proposed, preserving the general idea of simultaneous observation and regulation throughout.

The regulation of behavior extends not only to groupings of hardware and software, but also to those which comprise people. In business this form of regulation is called leadership, but in the military it is called command. This does not imply leadership to be unimportant in the military; on the contrary, military leadership is vital for victory. Nonetheless, orders are something unique to the military, and command a sign of decision making rather than conclusion forming or consensus building. Urgency and immediacy, characteristics of military conflict, call for unequivocal and instantaneous alignment with directives. In command, agreements with orders were secured upon joining; it does not have to be sought at issuance.

Apart from the very obvious role that systems thinking and systems practice has for the military,[22] the question we pose is: What is the significance of the concept of command outside of its military connotation and context?

For us it is less about the system and more about the system of systems (SoS) to which we have already referred. The word *command* suggests a commander and a commanded. But as Steve Johnson wonderfully points out in his book *Emergence*,[23] the colony of ants has no commander, only variously gifted (or designated) ants each following built-in rules of conduct that invoke relationships with neighboring ants. As a consequence, the colony, as a whole, develops patterns of behavior that protect the queen (the sole source of new ants), garner food, and bury the dead. This phenomenon and others like it, which Johnson beautifully portrays with examples of cities and software, are a challenge to the military style of command, one that leaders are taking most seriously, and to other complex systems in which there is no evident prescient commander to look to for direction and orders.

What makes this reinvention of the term so poignant is the fact that the military, consisting of several departments, exists in order to achieve victory, but this victory is now no longer in the hands of any one department. The various commands given to each circulate within but not necessarily across their respective system boundaries. The talk now is of "joint victory," "joint force," and "joint command," to which the original notion of command must now relate.

Not for no reason is the military thinking earnestly about network-centric operations and warfare where commanders spontaneously appear at the sharpest point of conflict with orders emanating from them to others who would normally expect to be directed by much higher authority. These same directives are good for other systems, each set up with a single purpose in mind, for example, to be part of a greater intelligence community, but none exactly designed to serve that greater purpose so that they fit in with all the others.

It is as though command has been inverted and been turned into "cooperative demand." By this we imply an urgent and immediate request from the heat of battle that conveys deep knowledge of a local situation coupled with a solid assurance that comrades with a view of the bigger picture can help and, in turn, help themselves secure the joint victory that all seek.

2.4 Time to Think

1. There are three houses (A, B, and C) and three utilities: gas (G), water (W), and electricity (E). Each house must get a direct, uninterrupted connection to each utility, but the various connections should not cross each other. Construct a diagram that shows how this is possible, or not, as the case may be. What does this exercise tell you about the notions of boundary, interior, and exterior? If you propose a nonplanar solution, how would the problem description need to change in order to restore the original constraint of connections not crossing each other? What learning can you apply to solving this new N-dimensional problem?

2. It appears obvious to locate a boundary around a corporation and to believe that its resources lie within and its competition lies outside. However, in the capital goods market, for example, gas turbine engines, some critical technologies lie outside the engine manufacturer, such as the engine control system, while the control system itself is embedded in the engine. Risk and revenue sharing partnerships can see the supplier become an integral part of the system integrator, that is, engine manufacturer, for many years, during which time he can learn much from the partnership and use this to his competitive advantage in working with other engine manufacturers. Likewise, companies that source products on the inside of others, for example, the CPU of a PC, can advertise to their advantage, for example, "Intel Inside," and give themselves an advantage outside of the company they supply to, an advantage they can use to invert the food chain. Discuss.

3. An old farmer dies, leaving his herd of cattle, seventeen cows, to his three sons. The will states that his firstborn should get half of the herd, the middle son is to receive exactly one third, and his youngest boy is left with one-ninth of the cows. The sons, who wanted to avoid fractional cows, could not figure a way out. One day, a neighbor, who was a lifelong friend of the old farmer and shared the father's love for the sons, came by to see how the orphans were doing. They told him

their problem. After thinking for a while, the neighbor said, "I'll be back!" He went away, and when he returned, the three sons were able to divide their inheritance in strict accordance with the will, without killing a single animal. How? What do you believe are the transcendental principles to glean from this situation, if any, that conceivably speak of faulty wholes, tricky parts, and sound relationships?
4. Consider Google, Inc., as a system. What are the inputs, outputs, and transformations of this system? What can Google do to improve each of these to its advantage? What might happen to these that Google does not have under its control that might do harm to this system? Take a piece of Google's history, for example, the purchase of YouTube. Were you able to predict this as an output, that is, purchase offer, based on the need for enhanced transformations? If so, how? What do you predict for Google today?
5. Consider a rural community, say in North Georgia, and the impact upon it of an influx of new residents made up of boomer retirees, foreign nationals, and, worst of all, "damned Yankees." Is the new variety good or bad for this community? How can harmony be maintained? What might parsimony mean in this contextual setting?
6. You are a screenwriter adapting a book for a film. The book tells the story of a young man born into slavery, sold to the owner of a school for gladiators, trained to kill his fellow graduates for the entertainment of sadistic onlookers, and finally hailed as the leader of a slave army that challenges the might of Rome's all-conquering legions. That man is Spartacus. In one scene he, like all his fellow trainees, is rewarded with the company of a woman, a female slave. He has never known intimacy, never been with a woman. They talk. He falls in love. Their dalliance is abbreviated by the realization they are being watched by his lascivious, unscrupulous owner and ruthless trainer. She is removed and assigned to another, a brutish Spaniard. Spartacus is left to himself, his thoughts, his hurts, and his concerns for a beautiful, tender, and intelligent young woman. In a later scene, the woman walks down the line of gladiators serving them food. No one is allowed to speak, on pain of death. She arrives at Spartacus. Their eyes lock in momentary reunion, his dangerously revealing an inexplicable devotion, hers an incredulous sympathetic response. He is confused and certain, strong and impotent, expressive and silenced. As an illustration of your power over parsimony, what line do you give Spartacus to capture this moment? Use no more than four words.
7. On September 11, 2001, months and possibly years of planning by Al Qaeda climaxed in inconceivable tragedy. Civilians living peaceably in the land of their enemy, trained to fly—but not land—planes by their enemy, using the trillion-dollar infrastructure their enemy had built, turned Boeings into bombs and sacrificed their own lives not to save

others but to kill thousands. What are the top three lessons relative to command, control, and communications that this event tells you?
8. In the story of David and Goliath, a young shepherd lad with an excellent track record for protecting his father's flock from the bear and the lion is pitted against a fully armed professional soldier, a giant of a man. David resists attempts to equip him with defensive armor for the fray, focusing entirely on offense and relying solely on his trusty slingshot and five smooth pebbles from the brook. Is this story in any sense relevant to today's terror-strewn world of asymmetric threats? What role is the United States being made to play? Is a change of role desirable or feasible? Can a Goliath win in this world? How? And what are the five pebbles that a David needs for victory, recognizing that in the Bible character's case only one was needed?

Endnotes

1. Churchill, W., *We Shall Fight on the Beaches*, House of Commons, June 4, 1940.
2. U.S. Declaration of Independence, U.S. Congress, National Archives, July 4, 1776.
3. Waldrop, M. M., *Complexity: The Emerging Science at the Edge of Order and Chaos*, Simon & Schuster, New York, 1992.
4. Weinberg, G. M., *An Introduction to General Systems Thinking*, Dorset House, New York, 2001.
5. Donne, J., *Devotions upon Emergent Occasions*, Folcroft Library, 1973, [Meditation XVII].
6. Ashkenas, R., et al., *The Boundaryless Organization: Breaking the Chains of Organization Structure*, Jossey-Bass, New York, 2002.
7. The eponymous creature in the movie *The Blob* (based on a story by Irving Millgate) could be said to be a system, however.
8. Reductionism is the principle tool of science that asserts explanations of phenomena are necessarily found by reducing the whole to its basic constituent parts, to the level of physics, and observing their behavior. If that is so, then why do chemistry and biology exist?
9. Simply Google *systemic failure* and observe the breadth of application of this term.
10. Hammer, J., and M. Champy, *Reengineering the Corporation: A Manifesto for Business Revolution*, HarperBusiness, New York, 1994.
11. Gharajedaghi, J., *Systems Thinking: Managing Chaos and Complexity: A Platform for Designing Business Architecture*, Butterworth Heinemann, Oxford, United Kingdom, 1999.
12. According to the article in the *New York Times*, October 31, 1971.
13. Churchman, C. W., *The Systems Approach*, Dell Books, New York, 1979.
14. Surowiecki, J., *The Wisdom of Crowds*, Random House, New York, 2004.
15. Resnick, M., *Turtles, Termites and Traffic Jams*, MIT Press, Boston, 1997.
16. Hock, D., *Birth of the Chaordic Age*, Berrett-Koehler, San Francisco, 1999.
17. Bronowski, J., "The Music of the Spheres," in *The Ascent of Man*, Little Brown, New York, 1976.
18. John 3:8 (King James Version).
19. Christensen, C. M., *The Innovator's Dilemma: The Revolutionary Book That Will Change the Way You Do Business*, Collins, New York, 2003.

20. "Keep It Simple, Stupid!" The principle roughly corresponds to Occam's razor and Albert Einstein's maxim that "everything should be made as simple as possible, but no simpler." Stephen, H., *Wharton on Making Decisions*, Wiley, New York, 2004, p. 137.
21. Senge was a Ph.D. student of Jay Forrester, the founder of industry dynamics, who foresaw the application of control engineering knowledge to business management, thereby making amends for the apparent failure of economists to unravel its problems. It is unsurprising, therefore, that Senge's seminal contribution to organizational learning should profit from control engineering principles.
22. The latest version of military language that we know of extends to the concatenation of seven critical subfunctions: command, control, communication, computing, intelligence, surveillance, and reconnaissance (C4ISR). One might ask, what does a generic C4ISR system look like?
23. Johnson, S., *Emergence: The Connected Lives of Ants, Brains, Cities, and Software*, Scribner, New York, 2001.

chapter three

Engineering

3.1 Pressed into Action

John Fitzgerald Kennedy, Ronald Wilson Reagan, and George Walker Bush may not have a great deal in common. Political convictions, sexual proclivities, and military service are some ways in which the lives of these men can be differentiated. One thing that they do share, holding the great office of president of the United States, means that regardless of character and conviction, they all labored equally under the same presidential pressures. That means each of them carried the awesome opportunity of winning the hearts and minds of people around the world to the values, responsibilities, and benefits of democratic freedoms. Using the resources at their disposal, each one in turn mobilized technological systems as devices to achieve the same political ends: that each and every individual could be persuaded of the virtues of democracy—to enjoy liberty, to live in peace and prosperity, and to fulfill personal goals and ambitions within a framework of social responsibility.

President Kennedy, realizing the advantage that the Soviets had gained in manned exploration of space, foresaw that unless the United States entered into a race with the communist-led superpower to land a man on the moon, as evidence of technological superiority and by implication sociological prowess, the battle for the imaginations of people everywhere might be lost, and with that battle conceivably the future of civilization.[1] In his own words to Congress, the charismatic leader of the free world said:[2]

> If we are to win the battle that is now going on around the world between freedom and tyranny, the dramatic achievements in space which occurred in recent weeks should have made clear to all of us, as did Sputnik in 1957, the impact of this adventure on the minds of men everywhere.… Now it is time to take longer strides; time for this nation to take a clearly leading role in space achievement, which in many ways may hold the key to our future on Earth.… Space is open to us now; and our eagerness to share its meaning is not governed by the efforts of others. We go into space because whatever mankind must undertake, free men must fully share.

Less than 3 months after President Kennedy's address to Congress and the nation, a wall was built that sharply and starkly divided the city of

Berlin, itself a landlocked hostage of communist control in a divided Germany, the outfall of the world's campaign to liberate Europe from Nazi tyranny. That wall remained in place for almost 30 years.

When President Reagan announced his Strategic Defense Initiative (SDI) in 1983, the Berlin wall showed no signs of crumbling. That which was slowly disintegrating, the Soviet Union and its impossible controls over the Eastern bloc, continued to throw a cloak of deception over its affairs, the wall acting as a symbolic iron curtain to disguise the truth. Reagan was taking no chances. Once again he challenged the nation, and in particular the science and technology community, to throw up its own impenetrable and invincible curtain, one that could intercept and destroy any strategic ballistic missiles before they reached U.S. soil or that of its allies.

President Kennedy's challenge—"to achieving the goal, before this decade is out, of landing a man on the moon and returning him safely to the Earth"—did more than launch a few rockets. It effectively launched the National Aeronautics and Space Administration. But to our minds, it did even more. It provided a new launch pad for systems engineering that we are privileged to describe in this chapter in a fresh, exciting, and accessible way. What Reagan's SDI program did, apart from inviting risible comparison with *Star Wars*, was to give us the term *system of systems*. Two decades later we are still eager to know what this means, how we will build them, and what new phenomena we will witness as they come into being and help shape our lives. We will touch upon these questions and that subject matter.

The forty-third president must have envied the opportunities his predecessors had in being able to address the nation at times where the danger lay ahead. When it came his turn to speak of threats against the United States, the world had already changed, and changed forever. In the sharpest of contrasts, our memories are now flooded with the unforgettable pictures of a silent and seated President Bush, in a Florida elementary school, listening incredulously not to a threat being made but to news of an attack that had happened. Terrorists turned Boeings into ballistic missiles and reduced towers into deathly rubble. Evil struck, turmoil reigned, death called, and hearts sank.

When the dust settled, one thing that emerged was the verity that systemic failures had been rife. What is more, these failures had occurred because systems themselves had failed as components of a greater system. People-centered systems had failed to work harmoniously together and with technology-based systems. This was true of the intelligence community in particular, which came in for yet more attacks regarding weapons of mass destruction, a crucial point of issue in the war on terror. People were asking questions like: Have we learned anything about systems? What more do we need to do to make systems safer, smarter, smoother? Can systems engineering help? Can it be more efficacious, more capable of dealing with highly complex systems, and more accessible to the professions that technology impacts, such as intelligence and policy formulation? These questions and that subject matter also figure in this book.

Three presidents in three eras faced with the same kind of threat, the overthrow of democratic freedoms, make similar choices—to enlist technology in the service of political persuasion and military might. Regrettably, the outcomes are far too unpleasantly comparable. Technology alone just does not cut it. It is especially galling when your enemy, the invisible terrorist cell, leverages the advanced technology you built, like the trillion dollar per annum telecommunications industry, to put its strategies into effect. An opponent that can turn your assets into liabilities forces you to reexamine the underlying assumptions, concepts, models, and methods that create these assets.

In writing this chapter we are resolved to scrutinize systems engineering from three vantage points. First, *systems thinking*, which we assert is fundamental to a restatement of systems engineering, being the wellspring of ideas and concepts for any revisions needful to updating the subject. Second, the notion of *system of systems* needs to be addressed, for we live in an age of complex systems that are continually, unpredictably, and fascinatingly interacting with each other, producing counterintuitive behaviors and emergent features that are altogether unexpected. Finally, we can no longer compartmentalize systems, be they people centric or technology driven, into commercial and military. There are now in existence only universal systems. Consequently, the phenomenon of the *extended enterprise*, rich in systems behavior, prevalent throughout commerce—especially the infotainment industry—and a key exemplar for system of systems, must become part of the systems thinker's mindset and the systems practitioner's case studies book.

3.2   The Way We Were

We are not sure when systems engineering (SE) began, but the term, some techniques, and its treatments of various problematiques have been around for more than half a century.[3] What do we have today that can be presented as the essence of this vital subject matter? What exactly do we have that can scale up? What is there about SE that is transferable to different classes of system? What can be adapted, by taking account of new problem situations or radically different opportunities that hitherto has been SE's domain of application? What is SE? What do people say it is? What did they tell Presidents Kennedy, Reagan, and Bush? The same tale or a different story?

In 1973 Barbra Streisand and Robert Redford, two great stars, graced a wonderful movie called *The Way We Were*. People with opposing political convictions formed a difficult but intriguing relationship. Systems engineering is supposed to be about parts and relationships and a way of forming a whole, even though the parts do not self-evidently have a relationship. We want to present our account of how well it does this and how badly. First, though, we want to present the essence of SE based on what has gone on so far: the way it is as things were then. We suggest seven essential ingredients.

3.2.1 Life Cycles

A top engineer who worked for Factron Schlumberger, a manufacturer of state-of-the-art electronic test equipment, once told us, "To be a test engineer you have to be bright enough to do the job and dumb enough to want to." He was saying in effect that the automatic test equipment (ATE) game was intellectually demanding, more high-tech than rocket science, but largely unglamorous since the draw for bright engineers was to work at the front end of new product development. The fact that new ATE products themselves required a design process, with all the front-end demands that poses, did little to obscure the ugly reality that it was really all about test and nobody much cares about that kind of back-end stuff. One thing that SE has done in essence is to find a way to put that care back by defining the *life cycle* as a governor of a product's (or service's) existence. In effect, a product has a story to tell because it travels on a journey, from conception through birth, adolescence, maturity, use, obsolescence, retirement, and death. Some products have a rebirth during their journey and others seem almost to be a reincarnation of earlier products.

The life cycle as a concept that has benefited greatly from the attention of SE is important for many reasons. It distinguishes important phases that then attract particular specialisms. Within any given phase there is opportunity for brightness to shine along with the more mundane. Phases can appear to act and behave independently, which is fine and proper, but they also belong to a parent who requires cooperation and interdependence. Some have chosen to build a life cycle within a life cycle phase extending the notion of parenthood. This ushers in the notion of organizational structures, a key means for ensuring that activities are executed in the correct order and with a timeliness that enhances quality in the eyes of the customer and minimizes nugatory effort and thereby cost. Organization never comes free, and due diligence is required to ensure that management attention is appropriate for any given product effort and to safeguard against seductive notions of recursion.

Life cycles themselves can have or need to conform to a life cycle. When a company chooses a particular life cycle, to manage its new product development process, for example, as Rolls-Royce chose to do with Derwent to manage the Trent family of gas turbine engines (see Figure 3.1a, b, and c), it can take time for this to become part of engineering and corporate culture. We remember that some parents are children, and so if life cycle is the box, there is always a need to look outside the box. For this reason, in time, the Derwent process, named after the first engine that Rolls-Royce built, became Create Customer Solutions within Rolls-Royce, a core business process that engaged the other business specialisms, not just engineering. This was a major development for that company, which historically has been engineering led. But when the industrial landscape changes, to reflect, say, prominence in marketing or contracting, life cycles have to change also.

Chapter three: Engineering 49

Create Customer Solutions

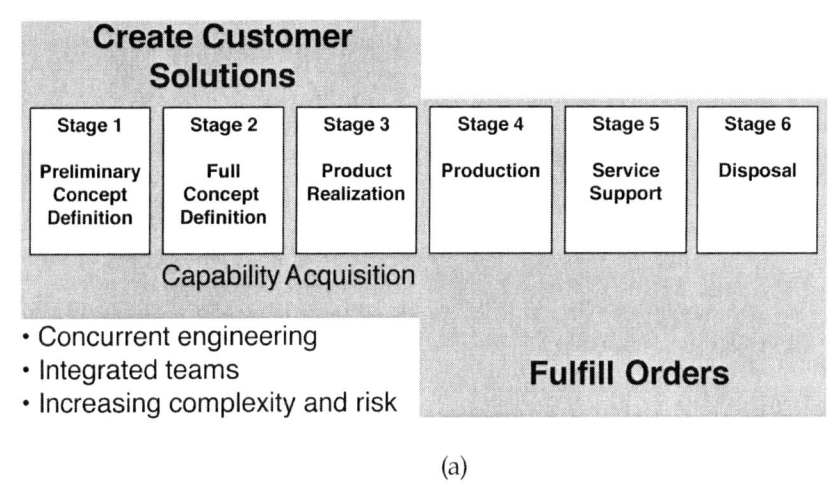

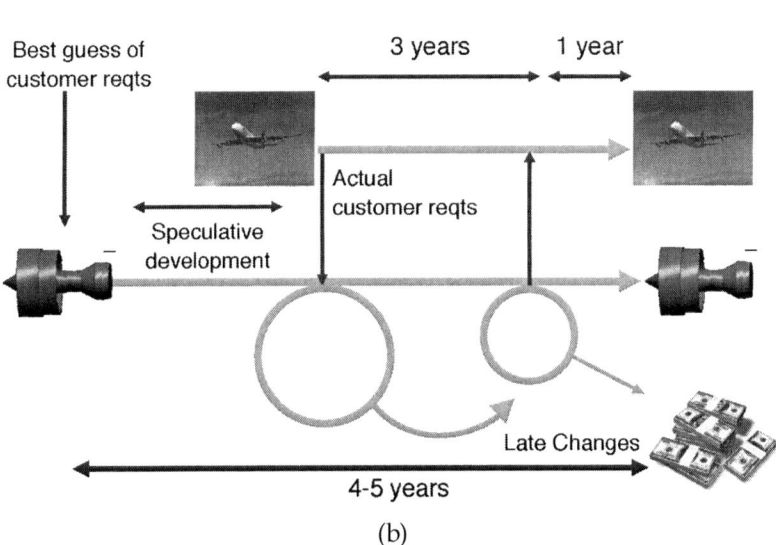

Stage 1	Stage 2	Stage 3	Stage 4	Stage 5	Stage 6
Preliminary Concept Definition	Full Concept Definition	Product Realization	Production	Service Support	Disposal

Capability Acquisition

- Concurrent engineering
- Integrated teams
- Increasing complexity and risk

Fulfill Orders

(a)

Best guess of customer reqts

3 years — 1 year

Speculative development

Actual customer reqts

Late Changes

4-5 years

(b)

Figure 3.1 a: Rolls-Royce business process architecture. b: Speculate to accumulate. c: Trent engine family development.

Finally, on life cycles, they are not merely unimodal operators, as you already may have noticed. True, their primary dimension is the temporal one, mapping phases in chronological order from cradle to grave. But there is also a contextual dimension when activity within the life cycle calls for finer-grain detail of whatever is being governed by the life cycle, for example, a new product, service, or organizational structure. Engineering is so much about analysis and synthesis, and this is reflected in the contextual dimension of the life cycle by the decomposition (and subsequent recomposition

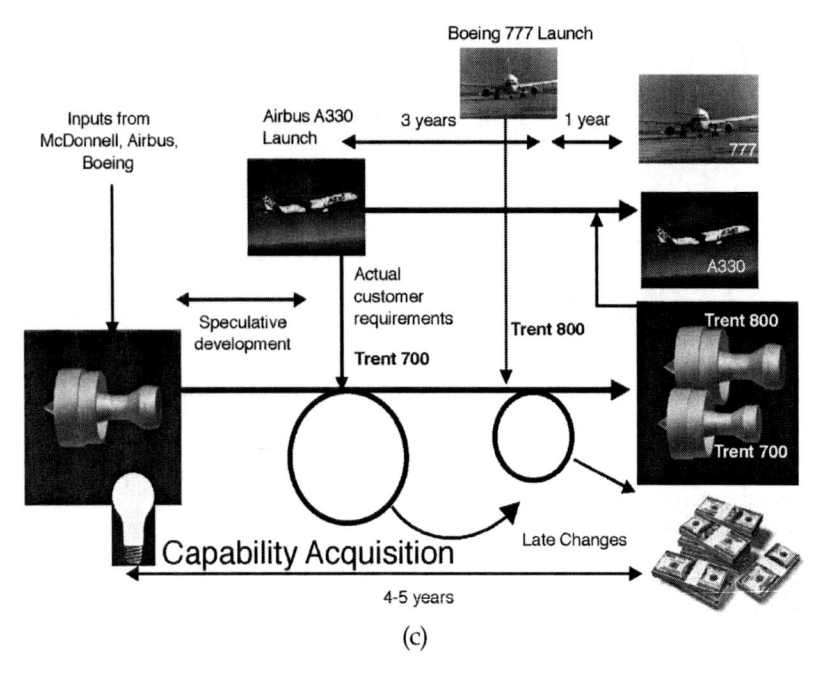

[Figure: Diagram showing Capability Acquisition timeline with Inputs from McDonnell, Airbus, Boeing; Speculative development; Airbus A330 Launch; 3 years; Boeing 777 Launch; 1 year; 777; A330; Actual customer requirements Trent 700; Trent 800; Trent 800; Trent 700; Late Changes; Capability Acquisition 4-5 years]

(c)

Figure 3.1 (continued)

or integration) of the thing that the life cycle governs. Likewise, there is a third dimension, what we might term the stakeholder dimension, which permits various interested parties to comment on or be actively involved in the unfolding of the life cycle. Agencies are scattered throughout life cycle phases and life cycle levels, but they can be congregated along this third dimension—life cycle views—and this gives rise to an interesting perspective. Agencies may have their very own life cycles for operating within a phase or at a level. The question now arises: Should these life cycles be harmonized or aligned in any way? And if so, how?

3.2.2 Passing Through

A second key concept, strongly related to life cycle, has been brought to prominence courtesy of SE, and that is the notion of *gates*, coupled with entry and exit criteria for passing through those gates. When Christian, the central character in John Bunyan's epic allegory *Pilgrim's Progress*, made his way from the City of Doom to the Celestial City he was strongly urged by Evangelist to pass through the Wicket Gate and proceed by no other means. Sadly, as so many of us often do, Christian ignored sound advice and got himself into dreadful trouble, not least of his troubles being in the Slough of Despond. When we are given a gate to pass through in order to get to some better place, through the gate we must go, eschewing shortcuts that at the time may seem pleasant and advantageous but which in time will lead us into even worse trouble than that from which we are fleeing.

Of course gates are not arbitrarily erected; they are the result, or at least should be, of considerable forethought and, what's more, earnest reflection of lessons learned from past, mostly failed, journeys. And the purpose they serve is to ensure controlled progress toward a final goal and to prepare the resources accordingly as further progress is envisioned. They become rendezvous points for sharing ideas, fulfilling targets, revealing information, receiving fresh direction, noting risks or hazards that may lie ahead in the systems journey, and taking advanced precautions to mitigate those risks. It is all commonsense stuff, to the point where it is difficult to see why SE should take any credit. But what is common sense now was yesterday a mystery or rank confusion. What is more, we know that many systems journeys today fail to respect these gates. People always know better, and sometimes shortcuts pay off—for them. Hardly ever do they pay off for others, and almost always it is the customer who pays up. This is making the customer more disconsolate, more regulatory, and more adversarial. Gates should be erected to make sense of progress from all angles.

Gates have signs above them. On one side the sign reads "Entry" and on the other it reads "Exit." To go through the gate you enter and then you exit. What exactly? In SE, the space you occupy as you pass through the gate is often called review. This term betokens an opportunity to review what has been done, to examine what exists, to sanction what is next proposed, and to establish specific criteria or qualifications for future review in addition to whatever is traditionally established for reviewing the next steps in the systems journey. The classical review points for SE are business requirements review (BRR), system requirements review (SRR), preliminary design review (PDR), critical design review (CDR), test requirements review (TRR), and production readiness review (PRR). These way points represent major milestones on the system journey, and the question in a lot of people's minds is: Are they fixed or can they be moved? In other words, are reviews schedule driven or event driven, meaning reviews should only be held when the event itself will be meaningful, that is, all the entrance criteria have been met. Plus, if exit criteria are strictly enforced, the review space is occupied for longer and the journey is temporarily suspended, making schedule-driven programs a less tenable proposition. Maintaining the balance between schedule and efficacy is part of the SE art and science, and the integrity of gates is fundamental to achieving this balance. One of our pals asserts: "Milestones should be fitted with castors!" While this is heresy for some, implying assent to slackness that inevitably leads to tardiness, for others it is wisdom, for who can know at the outset of a major program where the major milestones *must* be? And to adhere rigorously to a schedule can lead to delays, as we shall show. There exists a permanent inherent tension between the top-down program control setting major milestones in their respective places, and bottom-up reality in which the unforeseen (and unforeseeable) occurs, unpredictable behaviors emerge, and counterintuitive results appear. Undoubtedly both are needed, presenting us with a paradox, one of many that surface in our statement of systems thinking.

3.2.3 Eggs Is Eggs

The term *requirements* was attached as a label to all the review spaces we mentioned above. It should come as no surprise, therefore, to learn that we believe this concept to be the third piece of essence in the SE story so far.

Engineers are action oriented. That is their nature. And society should be thankful for this; otherwise, nothing would get done and we would all be stuck with what we have—not much. While this nature of wanting to get on with the job should not be considered unthoughtful, there exists an inevitable tension between execution and planning. The former gets things done; the latter apparently does not. The payback of making the latter an essential part of the former is to safeguard the execution process from nugatory action, which turns x into 3x. The cost of the former can appear to turn x into 2x. We grew up being told that "eggs is eggs," a corruption via dialect of the incontestable truth that $x = x$. The point is that x *never* equals x. Expending x always leads to additional expenditure, which in effect is the cost of interfacing action to planning, or thinking to doing.

Einstein's remark that "in the brain, thinking is doing" is helpful. It eliminates the inherent tension between execution and planning. However, it does not change the basic nature of the engineer. If systems engineering is not really engineering, nor part of engineering, but actually the servant of engineering (on which basis it is reasonable to include the word *engineering* in its description), then what is it that SE provides, by way of a service to action-oriented engineers, to ease the tension while taking nothing away from the splendidly unrivaled energies, passions, and expertise that engineers exhibit? In a word: requirements.

Requirements are part of the puzzle for tying the need, which impels the start of a journey (for example, "to win the hearts and minds of people the world over to the virtues of democratic freedoms over communism"), to the technological solutions whose realization, implementation, and deployment fulfill the need. Requirements are articulated by people, not always that well necessarily and not always by the technologically literate. Requirements are captured by systems engineers having the skill and knowledge to empathize with those struggling with the articulation and those who deeply understand the technological implications, ramifications, and limitations of interpreting what is captured. Requirements are a means of preventing technology push per se and of leaping to solutions before the problem (or opportunity) need has been grasped—an inevitable side effect of action orientation.

Requirements should not lead to paralysis by analysis, but sadly they do. We know of some complex SE projects that have birthed fifty-thousand-plus requirements (the so-called "shalls"). This we believe is a clear case of the problem parenting myriad problems that sabotage any birth of conceived solutions to the original problem. Requirements are bedeviled by ambiguity, simply because connecting people (as Nokia would have it) amounts to more than physical linkage by voice or data. None of us can really know what another means when we really do not know that other person. Uniformity of purpose and conduct, insisted by the military and formerly by Big Blue, goes part way to establishing context for communications. Imagine taking a flight from Orlando to Newark on Continental Airlines. The flight attendant announces: "Lower your window shades to obtain better viewing of the movie." Suppose you have seen the movie and it was awful; it would be better to watch it with the blinds up, so that the screen could not be viewed because of the reflective light. The lady meant "see better"; she did not intend "be better." How many SE requirements are prone to similar misinterpretation—with potentially drastic consequences?

Requirements are progenitive. But the midwifery business that serves them is incomplete in its competence. How do you unpack an idea? Sir Winston Churchill once said, "I am sorry to write you such a long letter; I did not have time to write a short one." He eludes to the complexity of abstraction and encapsulation. And so he should. Nor should we think that traveling in the opposite direction is easier. The business of amplification, enlargement, and expansion—in the semantic sense. Churchill, leading the fight against Nazi domination, was asked: "What is your goal?" He replied, "Liberate Europe." Great. Now how do you unpack that? Clearly, the requirements business has to be run by those who know what the words mean; more importantly, the people involved have to be honest enough to know when they do not know what they mean, and brave enough to say so. Requirements have become a major subject in the systems engineer's technology furniture store. That is not a problem so long as we realize that principally they are a cultural phenomenon. The best thing that this essence has achieved is to separate the problem space from the solution space. The skill of the systems engineer is to achieve this separation while simultaneously achieving its integration. In part that is about formalism and the technology that adorns this, but in large measure it is about comprehension, and that is a human thing. Speaking about people ...

3.2.4 See What I Mean

"We the people" is a great beginning, but then it gets tricky. Your freedom, my annoyance; your choice, my constraint; your cure, my poison; your East, my West. Sometimes you do not have to go to two to get conflict. My Lexus, but my olive tree (read Tom Friedman[4]). We want to improve, but we do not want to change. We want consistency, but we desire spontaneity. Make it better, but do not do anything. And the guy in the middle? The jolly old systems engineer. And his response to this variety? To identify, respect, and enlist all the relevant contributors to the fray, regardless of status but conscious of value. For this reason we suggest *perspectives* as the fourth essence of SE.

A recurring theme in systems thinking is the maintenance of differentiation and integration, of distinction and blend, of dispersal and togetherness, of variety and harmony, of personality and oneness, of parts and whole. This theme is the story of our own existence. Finding meaning in the confusion of detail, purpose in the thread of meaninglessness. How is it done?

We pose the question "What is a penitentiary, prison, jail?" For some it is a place of incarceration, ensuring the bad guys are kept locked up so that the good guys go unharmed. For others, it is a correctional facility, where the bad guys can learn that it is unwise and wrong to continue along the broad road to destruction, and instead discover the narrow way that leads to the good. Others will offer that jail is a place where proper punishment can be visited on wrongdoers; retribution is prime, with maybe redemption a second thought. Finally, some might observe that it is a university of crime, a unique opportunity for the bad guys to get better at being bad, where graduation leads to crime paying off. Who is right? And what gets built accordingly? How do you architect a place where this diversity of viewpoint exists. Are all wrong? Is one right? Which? Or are all right? And how can this be?

Some argue that the work of the systems engineer begins when this issue is settled—when the requirements are all gathered safely in. No one can blame anyone for wanting to wait until the confusion is at least defined. But we say that the systems guy has to become part of this confusion, without getting confused. He is in it and yet not in it. Another one of those ubiquitous paradoxes. How so? We assert that SE uses three key constructs for resolving this paradox. First, there is the notion of *stakeholder*, a person—not a company—who has a stake in systems thinking. Of course companies are permitted, except that they have to be represented by a unique individual. Achieving this "one from many" stakeholding is about as nontrivial as the case of aligning multiplicity of stakeholders. We have much to say about this construct in a future chapter.

The second construct is *viewpoint*. Each stakeholder will have at least one viewpoint. There will be a multiplicity of these, and the systems guy regards these as simultaneously tenable. In other words, they are all valid, they each have something to say, and they can each be "right," at least as the stakeholder who owns them sees them. Their "incorrectness" is challenged by the inconsistency they have with others, an inconsistency that may only be brought to light by available or foreseeable technology. At this stage in the journey, incorrectness is not regarded as a disaster but should be seen as an insight into future worlds.

The third and final construct is methodology; methodology is not method. Engineers love method. It is precise, unambiguous, deterministic—even when it is probabilistic and value-free. Systems people must learn to appreciate methodology, a set of principles, rules, heuristics, and tools (methods) loosely organized in ways that can be pragmatically tailored to unravel mysteries, resolve conflicts, and search out the feasible from the infeasible. We have been immensely impressed with the work many have put into creating a new methodology that directly addressed the softer issues in the systems journey such as culture, conflict, and collaboration. We applaud this move toward soft systems methodology, based on the virtues of what can now only be referred to as hard systems thinking and practice, the stuff that deals with the hardware and software methods and solutions in technology. What is more, we offer a contribution to this new movement that we believe facilitates tangible and traceable links between strategic intent and technological solutions: hand waving and box building.

3.2.5 Spoiled for Choice

At some point in the systems thinking process a fork appears in the road. What to do? Which way to go? Who to consult? How to evaluate? We suggest there are three principal ways to characterize forks. The classical fork relates to the product that an engineering team has to build. The basic question is: Which one of several candidates, sometimes referred to as conceptual designs (or quite inappropriately, in our view, system concepts), should we go for? It depends. A second fork is that when faced with a choice of process, or the technology means of achieving a transformation, possibly in product or service offering. A third fork relates to a selection of who—the individual, team, or firm—will execute a piece of the action. Determining a choice of fork is the stuff of decision theory that brings in risk analysis and other paraphernalia.

What is generic to all of these forks, be it a matter of what, how, or who, is that *candidates* can be identified, *criteria* formulated, *performances* weighed in the balance, and a *selection* made—rationally speaking, one that is predicated on the information obtained by application of the foregoing constructs. However, it does not always work out that way. The role of the systems guy is to provide as much service as possible to facilitate rational thinking and to report as rationally as possible when irrational decisions are made; after all, there may be some patterns of thinking that operate in choosing a fork in a journey that have escaped the received wisdom of the day.

The particular piece of systems engineering essence that helps the systems guy to fulfill this role, we suggest, is the *trade-off study* (TOS). The results of a TOS can be the subject of a review within the SE life cycle—a place where milestones and forks coincide. Alternatively, the need for a TOS can be determined by a review, making the TOS a key component in the next phase of the life cycle—the wagon train boss dispatching scouts to figure out future trails.

3.2.6 Modeling and Simulation

In a contest to choose the greatest understatement of all time, surely what would rank in the top ten is the line "Houston, we've had a problem." Commander Jim Lovell was simply stating a matter of fact. Turbulence apart, the

Odyssey's dashboard evidence was clear and unmistakable. It was not until much later that the full extent of the problem had been revealed, at which point Gene Kranz and his crew on the ground set about defining a solution—one that would bring Lovell and his companions, Fred Haise and Jack Swigert, at the time headed in the wrong direction, back to Earth alive and well.

Lovell tells his story in *Lost Moon*,[5] which became the basis of the movie *Apollo 13*. These works and stories give the briefest of clues as to how Kranz and company rescued the stricken astronauts. We have to look at *Moon Lander*[6] to dig deeper into how the problem-solving process unfolded, a process enabled by the sixth essence of SE: *modeling and simulation*.

Tom Kelly, the Grumman chief engineer with responsibility for building six Lunar Excursion Modules (LEMs), on a lecture tour he was making through Asia to explain the processes that underpinned his team's creation recounted: "The Grummies [sic] could not have turned the LEM into a lifeboat had it not been systems engineered." That gets your attention, which hugely reinforces ones respect for SE in general, and modeling and simulation in particular.

We have come a long way from simply making clay models of the things we plan to build, except of course these devices are still used extensively and valuably in the auto industry when it comes to new product development. Nevertheless, digital computing and mathematical analysis have led the way to more sophisticated models and the means of simulating them, to the point where yesterday's advanced simulators are hardly comparable with today's virtual-reality experiences that thrill theme park guests.

The *Apollo 13* movie shows Ken Mattingly, played extraordinarily well by Gary Sinese, repeatedly exercising the LEM simulator in order to configure its power budget for minimum sustenance of the astronauts while ensuring safe return of the service module to Earth. Simulators of this type, much propelled in sophistication by advanced technologies, are commonplace training tools nowadays. Aircraft pilots are qualified to fly passengers on jets for which they themselves have zero flight experience and are operating for real for the first time. What are the lessons from this spectacular development in prototyping technology, and what does the future hold for it, in the context of an evolving body of SE knowledge? We suggest five lessons and offer two pointers to the future.

First, we affirm that all models are wrong, some of them are useful. Since a model is an abstraction of reality, and that too only from a particular perspective, they are fundamentally wrong because they are *not reality*. That gives no license to models that are wrongly built—after all, two wrongs don't make a right. So usefulness, or purpose, is what determines a model's role, given that it is correctly formed. Models therefore have teleological value even though they are ontologically erroneous.

With this in mind, we would suggest, secondly, that models need not be of reality but can better serve reality by *reflecting ideality* and positing debate among executives who engage reality with the purpose of making systemically desirable and culturally feasible changes in and to reality. This is an important point that soft systems thinking leverages neatly.

Third, model creation should force us to scrutinize the relevance and significance of *perspective*. In that delightful movie *Pretty Woman*, Richard Gere's character (Edward Lewis) asks his escort her name. Julia Roberts (playing the hooker by the name of Vivian) retorts, "What do you want it to be?" When we make models we ask, "What do we want it to be (of or for)?" Too many models are made either because we can make them (we have a technology we love to use or want to try out) or because we are in a hurry and the first thing we see we model: the curse of action orientedness. In our view, *no* model should be built unless we know what we are looking at, why we are looking at it, from where (which standpoint) we are looking at, and what it is we believe we can see better because we will have the model. The last thing of significance in building a model, in our consideration, is the *how*.

For this reason we assert that a *family of models* (risking the charge of paralysis by analysis) is a desirable goal, especially when a variety of models—digital, analog, iconic, metaphoric—can be deployed in order to enrich the utility factor and so accomplish our goal, for example, to pass through a gate, to conduct a trade-off study, or to interpret requirements.

Our fifth and final observation relates to the *adaptability* of models as they convolve into a higher-order model. A maxim of systems thinking is that a part cannot be removed from a system and be the same. In other words, the part becomes something different when it truly belongs to a system. But what if the part is a model and the system a collection (or family) of models integrated together? Do we observe that this model changes? Structurally it remains the same, though dynamically it changes its behavior depending on its interactions with other models and the state of the system. But perhaps a model, as a part, should change its structure, and in ways that we cannot foreknow.

As we contemplate the marvels of modeling and simulation technology we cannot help but observe that the future will be driven by people, billions of them, and their essential needs. It is not for people to become something that they are not in order to take advantage of technology. That they may do, but not of necessity. It is for technology to adapt its form in order to serve people as they are, where they are, and what they choose to become and do. We know that this is happening, and indeed technology has journeyed along this road. What more do we see happening? The plain fact of the matter is that people do not see product, or processes, or the enterprises that provide them. They see what happens to their lives when they try to use products—for good or ill. They experience time delays, poor quality, broken promises—the dreadful services we are all used to meeting—and an explanation of internal processes palliates this experience not one whit. And people encounter people not firms. First the front line and thereafter the back office, and usually with little joy. The challenge for the systems guys, and it is an immense one, is to gravitate the products, processes, and enterprises to the point where consumers find their lives enhanced by products, fulfilled by services, and enriched by encounters. If this be so, them modeling and simulation in the future should serve this purpose.

What happens of course is that this agenda is judged to be a pursuit of the Holy Grail, and therefore infeasible. So the modelers do what they can do in order not to be unoccupied and therefore unuseful! We say, let us see if it can be done; after all, it is what people want, and they are the customers. To start, we call for a simultaneity in modeling whereby product, process, and enterprise are consistently modeled as one (whole system) with component and adaptive models of each integrated together. This call, though challenging, is feasible, and undoubtedly desirable. We touch upon this in Chapter 3 and Chapter 7.

That is our challenge to the modeling community. What of the simulation community? The preoccupation with all firms is to demonstrate the strength of what they deliver, be this to another firm or to the end user or customer. This is perfectly understandable, but its importance is marginalized by the *how* of that delivery. What do we mean? It is becoming increasingly clear that what gets into the customer's lap is a consequence of myriad firms combining in some way or other. We say *combining* because we do not believe this behavior is deserving of the term *integrating*. That is our point. What if the end user, or her representative, insisted on validation of this combining operation, this integration, ahead of any product or service delivery? This is something that can surely start with the capital good market, especially the larger ones like Future Combat System (FCS).[7] What if the U.S. government insisted upon a contracting base demonstrating its competency to deliver a technology or family of technologies on an on-time, on-cost, on-quality basis. Of course this base is a family, a dynamic one we grant you, but can it behave like a family, in which squabbles are sorted and parental control directed at maturing children?

We believe that simulating the product (or service) has had a fair crack of the whip. It is time for visibility into the black box to convey confidence that what will emerge will be what people really need. We call this competence demonstration or enterprise realization assurance, as opposed to technology demonstration. And we believe that the underlying technology and methodology (management simulation) is already robust enough to permit its extension to this challenging call. This we will talk about further in Part 3.

3.2.7 The Long Haul

While engineers, by nature action oriented, love to make a fast start and get on with the job at hand, they typically do not stay with a project for its duration; there is always another fresh start to be made, infinitely more attractive to an innovator than seeing through the old idea. Some engineers we know have worked their whole careers, 35 years or more, on a handful of projects, seeing each one through to a reasonable completion. But these we find to be an exception. It is far more likely for an engineer nowadays to work on one-hundred-plus projects serving an average of less than 3 months on each. This might or might not be a problem. What is a problem, however, is to have a lack of vision of the entirety of a project, a myopia that limits rationale at the front end and causes grief and possibly catastrophic failure at the back end.

Our final piece of essence in the SE makeup is *operational effectiveness*, a term that restores the role of long-term vision for a project at its beginnings without necessarily insisting the engineers devote their entire lives to seeing the project through to the end. In particular what the systems professional needs to be aware of when regarding operational effectiveness is the need to architect sound foundations for the project, and to use design tools that reach into the project's future well-being.

The foundations to which we refer have been built into a graduate and executive education program at Stevens Institute of Technology.[8,9] As such they are both a model of how long-term vision can be sustained in addition to being a dissemination means to upcoming generations of system engineers who can be expected to replicate these foundations in future systems projects. The architecture of these foundations comes from unpacking the notion that the role of systems engineering is to provide a high-quality product that will serve the interests of the customer in terms of both technical and business needs. Instantly, this conveys notions of profitability, of total system effectiveness, of total ownership costs—not just acquisition and deployment, and longevity. In other words, systems engineering serves the customer in the long haul, not just the quick fix, rapid response, and all other short-termisms that we know of.

The constitution for these foundations is characterized in the schematic shown in Figure 3.2.[10] The first thing that interests the systems engineer is to endow the system he designs or architects with the capabilities and characteristics that lead to high performance. For this he develops from the

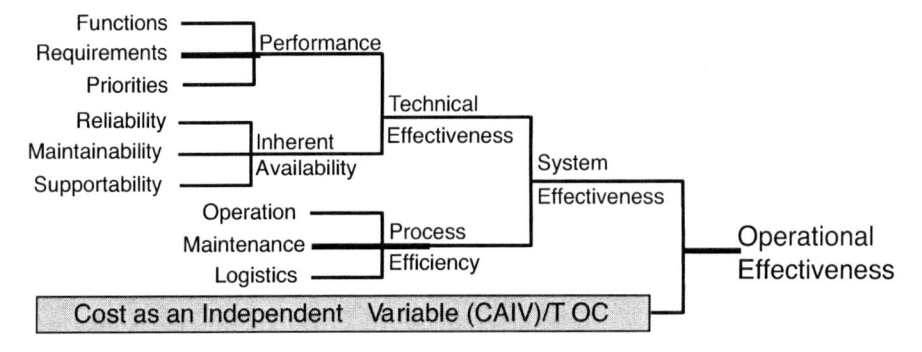

Figure 3.2 System operational effectiveness model. (From Verma, D., J. Farr, and L. H. Johannesen, "System Training Metrics and Measures: A Key Operational Effectiveness Imperative," *J. Sys. Eng.*, 6, 4, 2003. Reprinted with permission of John Wiley & Sons.)

various perspectives he receives both functional requirements (what the system needs to do) and nonfunctional requirements (what the system needs to be) and priorities that will determine relative importance of these requirements, which inform trade-off studies when system concepts or candidates are posited. But this merely takes care of performance; it does not take care of availability, and as common sense tells us, an unavailable system capable of high performance is of little value. Therefore, the systems engineer must ensure availability by detailed consideration of reliability, maintainability, and supportability, at the outset of his efforts, making these considerations part of the foundations of his systems practice.

Is this the end of the matter? Actually not. Performance and availability combine to give technical effectiveness. We need also to include process effectiveness, the habits of the system when it is operated, maintained, and energized by the logistics that are needed. If the system is a rifle, it is tempting to let the system boundary be confined by the space the rifle occupies. But what about the bullets? What about the training of the soldier for whom it is intended? What about getting the soldier to the scene where the rifle is to be unloaded—at targets. No one wants to suffer the curse of the ever-increasing system boundary. But by the same token, an overly restricted system boundary will render any attention given to the system within it largely ineffective. Systems are inevitably embedded within wider systems, but the notion of operational effectiveness respects this and provides a controlled means for expanding the system boundary, rendering effort expended within each valuable to outer layers.

The combination of process effectiveness and technical effectiveness leads to system effectiveness. Is this the end of the matter? Regrettably not. But there is only one more step to take, and that is to consider cost as an independent variable (CAIV), which combined with system effectiveness gives the long-haul view for the systems engineer, which is true operational effectiveness that includes the notion of profitability. In the military world, CAIV is a key strategy for reducing total ownership costs (TOCs). It relies on two principles: (1) system cost is constrained and (2) trade space is the venue for making smart decisions. Trade space is commonly defined by alternatives in terms of the performance, cost, and schedule impacts that each alternative presents. Risk must also be included in two ways. First, risk is a fourth dimension in the trade space, recognizing that critical decisions may be driven by the risks of certain alternatives. Second, risk actually "discounts" the anticipated performance, cost, and schedule options; in other words, it lessens the trade space to ensure a decision maker does not trade away something that may not be attainable. For example, assume you have an aircraft with an anticipated range of 12,000 miles versus a requirements threshold of 9,500 miles. You could trade away up to 2,500 miles of range for a fully tested, validated system and still meet threshold. However, you would not trade away 2,500 miles of range at the beginning of program definition and risk reduction (PDRR), when there are potential weight growths, fuel consumption increases, and other parametric uncertainties.

3.3 Quite Another Story

In our book we have a great opportunity to present the reader with an account of systems thinking that is true to its past, accessible to a whole new readership in the present, and relevant to future technology worlds. We have half a century of education, research, and practice between us as writers, instructors, consultants, innovators, and engineers. We are placing all of this at our readers' service in line with our doctrine that systems thinking is a servant. We truly believe that to lead is to serve, that to be great is to be humble, that strength comes from meekness, and that intelligence is greater for community. Our beliefs and our values enable us to present a faithful, accessible, and, most excitingly, refreshingly inspirational account of systems thinking. An account that uses story as metaphor, that exploits paradox as leverage for meaning, that treats journey as experience and vice versa, and that reinterprets the essence of past contributions as signposts to true systems thinking.

Everything about the essential ingredients of systems engineering we have presented above points to the use of journey as a metaphor. The life cycle notions are the ages of a story or the phases of a journey. The review gates are major milestones or way points that enable a time of reflection of the past, of summation in the present, and of preparation for the future. Requirements present themselves to us as either signposts or landmarks for the way forward, and compositely a map of what the terrain looks like, what features we need to take careful note of, and what boundaries to respect. Requirements also appear to us as essential resources we carry with us on our journey, some of which are a direct resource and others a means of deriving subsidiary resources just as matches, lighter fluid, and kindling are the primary elements of fire starting.

Perspectives, in the guise of stakeholders and their various viewpoints, are characters we meet on a journey. Sometimes these are good companions whose advice, encouragement, and support we highly value; sometimes they act as Job's comforters, knowing what is right and good for us, yet giving themselves to us in an awkward, negative, and unaccommodating manner. Perspectives are most unhelpful, as are some companions, when they are maliciously compliant and grossly disingenuous; they are there for a reason but not always for what is apparent. Choices are the stuff of journeys; there are choices of when to stop and rest, and when to resume; choices as to which way to travel and by what means, at what speed with what risk; choices of whom should go with us, at what stages of the journey, and whom should be left behind. Modeling and simulation are to us the scouts who reconnoiter the terrain ahead of us; models are the vehicles for the scouts and simulation is the news they bring back. Sometimes it is news that the land is indeed flowing with milk and honey, but the people who live there are powerful and the cities fortified and very large (negative recommendation). Sometimes the same exploration but with different scouts yields an opposite recommendation: "We should go up and take possession of the land, for we can surely do it."[11]

Finally, what journey worth making is not of the long-haul kind in which the character of the journeyers is properly developed? Who looks at a map, figures out a route, obtains all the resources needed for the trip, and then stops at the front door? Systems engineering is for the long haul. It defines the journey from start to finish and serves all who make it for the entire duration, not just the early phases. Systems engineering is a journey that takes the full 20 years (or any other number you care to choose), and is not simply a twenty times repetition of the same, more or less, 12-month trip.

We assert that systems engineering is a process—a process that transforms a functional need, a mission capability requirement, or market opportunity into a complete description for a system that meets the need. We have created a road map for that process that delineates the steps that systems engineers take in performing this translation (see Figure 3.3).[12] While repeatedly emphasizing the iterative (at the same level) and recursive (at different levels) nature of the systems engineering process, we use the road map to structure the flow of systems engineering activities, from determining stakeholders and stakeholder requirements to generating, evaluating, and selecting concepts, developing the operational view, and ultimately producing a complete system architecture that is testable and verifiable. Certain key milestones and concepts have been developed to ensure that the project and the system developed remain true to the original intent.

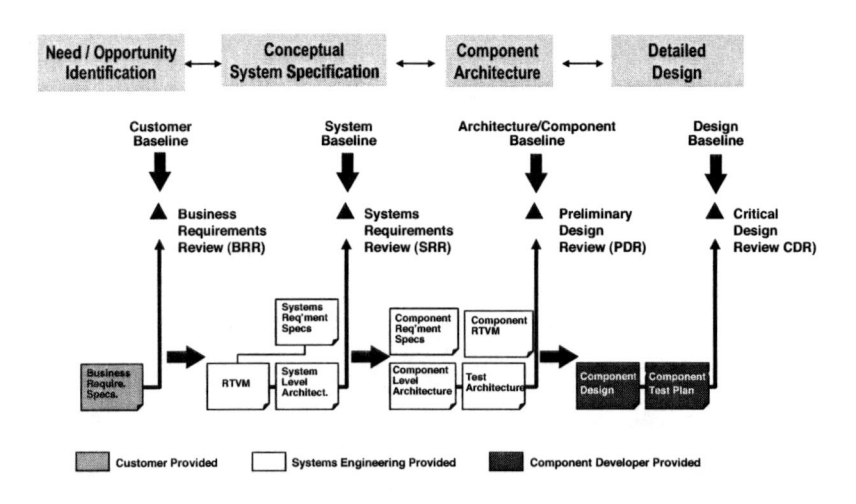

Figure 3.3 Systems design and operational effectiveness systems engineering process.

But the systems engineering process is far more than just a sequence of steps—it is a thought process. Indeed, the thought process is far more important than the sequence of steps. A person who thinks like a systems engineer can produce excellent results following a variety of different paths. Conversely, a person who does not understand the systems engineering thought process can rigorously follow a prescribed sequence of steps, yet produce output of little or even negative value. In order to present an accessible and inspirational account of systems engineering, it is essential that we present not just the steps to be followed but the way a good systems engineer thinks as he or she executes those steps. This thought process provides the necessary robustness and flexibility to the systems engineering process and its tailoring to suit the specific circumstances of the problem or opportunity being addressed.

For us this thought process is a journey—the systems journey. And it is not just about thinking, it is about doing while thinking. The systems journey is systems thinking manifested in systems practice, and it takes all those who will go, on an exciting journey, one outcome of which is new technology worlds. But other outcomes, equally important as far as we are concerned, are the experiences of the journey, the stories that can be told on the journey, and the accumulated wisdom that can increasingly make successive journeys more rewarding for recipients and more enriching for sojourners.

In presenting this book to you we will accompany, wherever appropriate and never frivolously, sound technical material with vignettes from our lives and snapshots of great stories, told in poems, books, music, and movies. Favorites for us, possibly familiar to you, will help your journey be a pleasant one. It is more than a collage but less than an epic; that is our work for you. However, take the systems journey and you will, we believe, encounter an epic. You will confront paradox throughout, and having been mystified, you will get enlightenment. You will get dragged into plots and emerge with explanations. You will be surrounded by characters and deafened by their clamor for attention to what each knows to be the truth; then calmly and serenely you will recognize a still small voice to direct your steps. You will appear to go round in circles but then move forward, realizing that diversions, iterations, and regressions are all part of progress. Enjoy!

3.4 Time to Think

1. President Kennedy's challenge to the nation came with a fixed schedule, a generous budget, and a clear and nonnegotiable goal. Systems engineering is often characterized as a navigation mechanism for steering a course defined by time, cost, and quality budgets. People generally agree that you can fix any two of these but not all three. Hence, SE is the means of getting the best of the rest. In your view, what are the principal sociopolitical challenges that confront us today and which budget—of time, cost, and quality—is the most important to fix and to

flex in addressing these? Describe three such challenges in less than five hundred words each, and amplify one of these using no more than twenty-five hundred words.
2. How migratable is the SE process to other domains of human endeavor, for example, the design and maintenance of an enterprise or people system such as the cast and crew for making a movie?
3. Engineers are at their best being creative and innovative as opposed to following procedures and consulting checklists. Consider the successful transformation of the lunar module into a lifeboat and compare and contrast the roles of creativity and conformance (to protocols and policies) in the contribution to that mission's success.
4. The execution of the SE process is inevitably one that gives rise to tensions. For example, that between the top-down program control setting major milestones in their respective places, and bottom-up reality in which the unforeseen and unforeseeable occurs, unpredictable behaviors emerge, and counterintuitive results appear. What, in your opinion, are the top five heuristics that engineers need to know and practice, relative to addressing the tensions that arise when temptations seduce us to take shortcuts, ignore document readiness, exit a review prematurely, or downgrade risks?

Endnotes

1. Schefter, J., *The Race, Century*, New York, 1999.
2. Kennedy, J. F., Man on the moon, Special message from the president to Congress on urgent national needs, May 25, 1961.
3. Classical texts, e.g., Chestnut, Daenzer, Hall, Machol, Quade, plus later treatments by Fabrycky and others.
4. Friedman, T., *The World Is Flat: A Brief History of the Twenty-First Century*, Picador, New York, 2006.
5. Lovell, J., *Lost Moon: The Perilous Voyage of Apollo 13*, Houghton Mifflin, New York, 1994.
6. Kelly, T., *Moon Lander: How We Developed the Apollo Lunar Module*, Smithsonian, Washington DC, 2001.
7. See www.army.mil/fcs.
8. Verma, D., J. Farr, and L. H. Johannesen, "System Training Metrics and Measures: A Key Operational Effectiveness Imperative," *J. Sys. Eng.*, 6, 4, 2003.
9. Verma, D., and B. Gallois, "Graduate Program in System Design and Operational Effectiveness (SDOE): Interface between Developers/Providers, and Users/Consumers," paper presented at the International Conference on Engineering Design (ICED), Glasgow, United Kingdom, August 2001.
10. Verma, D., et al., *J. Sys. Eng.*
11. According to the Bible (Numbers 14) the report from some of the spies Moses dispatched to explore Canaan was negative; a lone voice, Caleb, dissented. Moses went with the majority and so Israel wandered in the desert for 40 years, 1 year for each of the days the spies had been at their work.
12. SDOE Program at Stevens Institute of Technology, www.stevens.edu/sdoe.

chapter four

Dynamics

A basic question that anybody might ask of any system is: How is it doing? How is it performing? Is it doing as well as we wish, or could it do better? It does not matter whether the system is a sales department of a mail-order PC company or the company itself. It does not matter whether it is a high-rise office building or one of the elevators ferrying workers between stations. It can be the engine in an automobile, a roller lifter valve in the engine, or the highway on which the car journeys. "How is the system doing?" is an inescapable inquiry.

Now the answer to this question might be based on opinion and, people being who they are, opinions are pretty unstoppable. Nevertheless, it is always possible that the answer can be more scientific; it can be based on facts, on measurements, on a formality that neutralizes bias, informs opinion, and rationally leads to corrective action.

Of course the system may not be doing at all well. It may be in a disastrous state. For example, the sales department is depleted because a competitor offers better bonuses; the elevator catches fire; or the highway is gridlocked. These snapshots of system performance, events in time, relay pertinent and urgent information on system performance, making corrective action imperative and largely unambiguous. However, such events may not tell the whole story. How is it that a competitor can offer superior remuneration? Is it because the company losing sales staff is creaming off too much profit, or losing revenue because of lack of customer service? Is there a fundamental fault in the elevator design that makes it a fire hazard? And how often do gridlocks occur? Do we have too many vehicles on the highways? Or not enough bandwidth? Or inadequate smart technology that alerts drivers in time to make alternative plans? Events are not unimportant. But they do not always tell the whole story, just like the system does not perform in isolation. It is part of a greater piece. And has behavior over time, turning events into patterns, making patterns a sounder basis for answering the question "How is the system doing?"

4.1 Thinks Can Only Get Better
4.1.1 A Systems Language

Peter Senge was mentored by Jay Forrester, the acknowledged father of industry dynamics, and that tutelage was not in vain. *Fifth Discipline*[1] became a best-seller, propelling Dr. Senge to fame and fortune, and in the process systems thinking (aka the fifth discipline of the learning organization, the others being personal mastery, mental models, building shared vision, and team learning) became more widely known and was made more accessible to the world of work, especially to those with an interest in management. Forrester's interest lay in creating a framework for formal comprehension of corporate dynamics, taking the body of control systems knowledge as the paradigm for capturing interactions, interdependencies, and feedback mechanisms in the world of work. Peter Senge's brilliant insight was to make available a systems language for answering the simple question "How is the system doing?" He knew that by keeping the grammar simple, sharpening the vocabulary by domain knowledge, and providing a graphically elegant communication medium, thinking about the system *and* the meaning of system queries would converge into a powerful tool to formulate effective action to create efficiency. Senge and the movement he inspired give us the means to develop structures with which to depict system behavior that can explain events and patterns. It is worth a look.

4.1.2 Servers and Clients

A system, it is said, is a collection of parts together with their relationships that forms a whole that serves a purpose that is meaningful to the system alone, that is, not to its parts or their relationships. Peter Senge has created a system, this being a systems language, that serves the purpose of creating other systems, these being models of reality fittingly described in his systems language. What are the parts and relationships of his systems language system? The parts are called variables, names carefully chosen and strongly identifiable with the domain being described, that is, subject to change in value over time. We are going to borrow an example from the excellent tutorial text by Anderson and Johnson, *Systems Thinking Basics*[2], to illustrate.

ComputeFast[3] is a leading provider of mail-order PCs in the United States. By combining low production costs, a customer base of small businesses and technically knowledgeable users, and a "no frills" corporate style, ComputeFast is able to undercut the prices of competitors, and thereby in its first decade of life has seen revenues grow from $100,000 to over $1 billion. Its forecast growth over the next 2 years is to $1.7 billion, and this is to be achieved by penetrating new markets, including overseas, with a new production facility being established in the Far East. This takes place against a background of a first-ever slump in sales over the past 12 months that alerts management to the possibility of declining customer service quality. Efficiency is the name of ComputeFast's game, and now the company has an event to consider (sales slump), patterns to reflect upon (downward trend in customer care), and a systemic structure to compose (its own industrial dynamic). Can systems thinking help?

First we need to identify some key variables that capture ComputeFast's dynamic. The naming of these variables is key, since they will form part of a language description, and semantic continuity and fidelity are important

Chapter four: Dynamics

considerations for comprehending the dynamics. Sales revenue is a simple but important variable, and two others relate to market expansion and customer demand. How do these variables interrelate? One way of depicting this is in Figure 4.1. This shows the variables as nodes or vertices, and the relationships between them are links or arcs that are directed, that is, the one variable is acting as a cause by influence and the other gives rise to an effect; hence, this type of diagram is known as a causal loop diagram (CLD).

There are two types of relationship: reinforcing or balancing. In the former a rise in the one variable leads to an accompanying rise in the associated variable. Similarly, if there is a fall, the latter follows the former. Thus, this relationship is labeled S for same. On the other hand, a rise in one variable leads to a fall in the other, and a fall leads to a rise. This type of relationship is labeled O for opposite. By following a loop, ending up at the first variable after several influences on other variables, we need to determine whether this loop overall is a reinforcing loop or a balancing loop. If the former, then this can be either virtuous or vicious, depending on the point of view. Either way it demonstrates positive feedback. If the latter, then the overall effect is a balancing one, demonstrating negative feedback.

A simple rule for determining whether a loop is reinforcing (R) or balancing (B) is to count the number of O's; if this is an even (or zero) number, then the loop is reinforcing. However, it is always good policy to check the semantics of the loop to see if this is true or not. In the example below, the sense of it is that as ComputeFast expands its market presence, its attractiveness to customers increases and their demands for ComputeFast's products similarly increase. This leads to an increase in sales revenue, which positively fuels the market expansion. This diagram does not show any limits to this virtuous growth, although there must be some. More customers mean more revenue, but they carry greater demands for technical support, and if this does not increase, then the quality of that support will decrease.

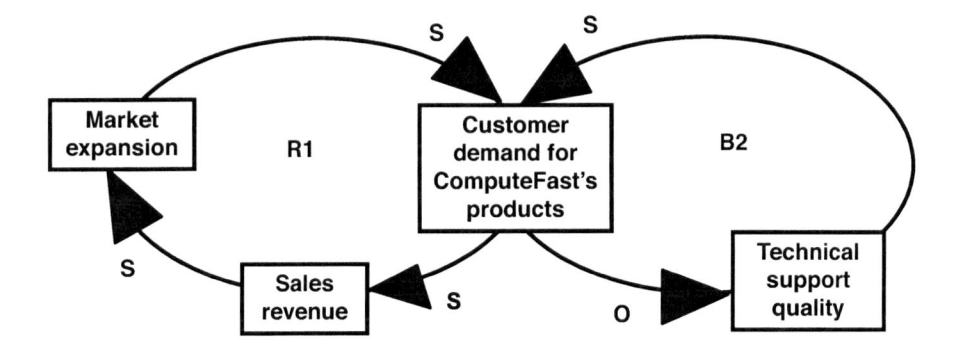

Figure 4.1 Balancing demand.

In Figure 4.2 we show this by indicating that these variables, customer demand and technical support quality, have an opposite relationship. However, technical support quality and customer demand have a similar relationship. If you look after your customers, they will be happy, stay with you, and attract more. Conversely, if you tick them off, they will be disappointed, may look elsewhere, and may say bad things about you. A happy customer tells his friends; a miserable customer tells everyone! This new loop is a balancing one and limits the growth of the first loop. However, what is the overall effect? If customers leave, then the sales revenue falls off; this halts market expansion and places less load on technical support so quality should be restored and the fortunes of ComputeFast will cycle. But how does it do this? What is the pattern? And how can management make decisions to ensure stability and even gradual overall improvement? More to the point: How can systems thinking be of use here?

Clearly ComputeFast will need to monitor the technical support quality, and in doing so will probably want to compare it to predefined standards. That comparison produces what we might call the quality gap, which is negatively influenced by the technical support quality; that is, if the support is there, the gap is reduced, and if it is not, the gap increases. This gap can be used to trigger investments in technical support capacity that, after some delay, will deliver the greater support capacity, and this positively influences technical support quality. That loop is a balancing one. However, the two

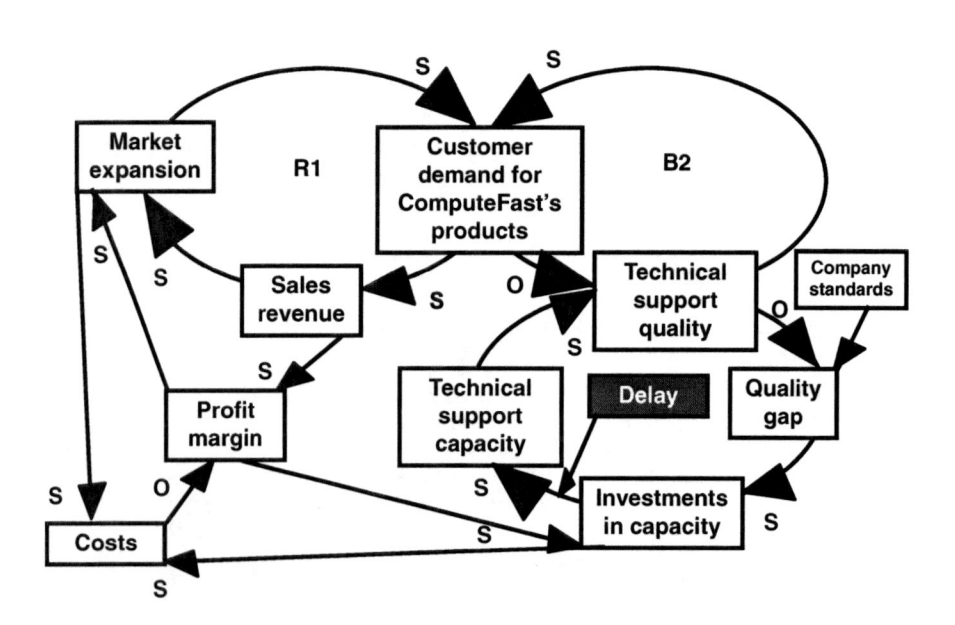

Figure 4.2 Simultaneous goals.

Chapter four: Dynamics 69

balancing loops create a figure 8, which can have a most deleterious effect on ComputeFast's well-being. For instance, as technical support quality increases, the quality gap is reduced, requiring less capacity for technical support, which will eventually reduce technical support quality; in the meanwhile, it is increasing and driving customer demand, although the capacity to meet that demand is not keeping up. What actually happens depends on the values of these variables including the important matter of delay. What systems thinking does is to lay this phenomenon bare, even though it may already be widely appreciated by those locally affected. However, what this type of language does is to create the opportunity for debate among a wider audience ensuring that action is not merely event responsive and not even pattern based; rather, it is predicated on a deeper structural understanding of industry dynamics.

One final point before we leave ComputeFast to its management: investing in technical support, to increase quality and maintain customers, is in a real sense at the expense of market penetration, which is also about increasing the customer base. ComputeFast's question must be: How do we achieve a correct balance? Moreover, both measures drive up ComputeFast's costs, which erodes the profit margin, which limits the drive to increase market presence and technical support quality. The picture, though still comprehensible, is becoming rather more complicated, and in reality the dynamics are indeed complex.

4.1.3   *Archetypes*

An important legacy of Peter Senge's work is the repertoire of system archetypes that pattern match numerous real-life situations, making these more attractive for treatment by the systems language as well as breeding confidence that the practice of this language, for efficiency purposes, will pay dividends. We will look at five of these. The first is known as fixes that fail. What this does is to look at how a given solution addresses the problem situation not only directly but also through unintended consequences. As Figure 4.3 shows, the direct linkage forms a balancing loop, whereas the existence of unintended consequences introduces a reinforcing loop. The existence of a delay in that loop can lead to an extended cycle of symptoms, fixes, and problem recurrence. To illustrate: a company may opt to downsize in order to improve profits, and sure enough in the short to medium term this fix does the trick. Often the easiest people to let go are the older folk. But they are also the ones with a lot of experience and knowledge whose loss is later felt via poor productivity, an unintended consequence that is also affected by loss of morale brought on by downsizing and the obvious feeling that "in time this is what will happen to me." As well as this being a detection mechanism, it can also be used preventatively in the sense that in taking any kind of efficiency action one might ask, "And what are the unintended consequences?"

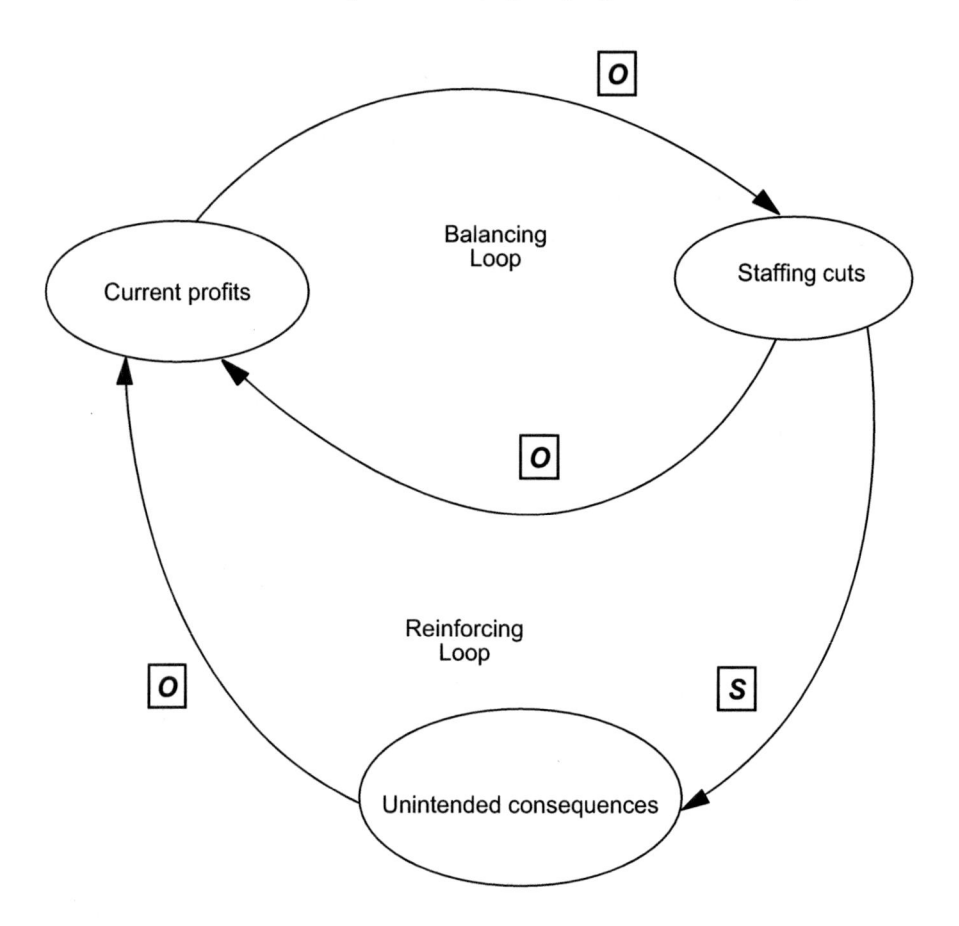

Figure 4.3 Fixes that fail.

A second archetype is one we have already seen in the ComputeFast example. It is called limits to growth. What happens here is that an action leads to success, increasing the demand for more action; that loop is reinforcing and is all about growth. However, a corollary of the success is a limiting factor that over time balances out the success formerly achieved. Thus, market expansion produces customer demand, increasing revenue and stimulating further expansion. But the greater demand has a side effect of reducing quality of service, which dampens demand. The lesson here is that going for growth requires an inspection of the unseen balancing loops that are going to limit that growth so that more measured action can be designed to improve efficiency overall.

Another archetype is known as shifting the burden, and we have personal experiences to illustrate his. One of us had the inestimable pleasure of consulting to Rolls-Royce on a project to dramatically reduce the time to

Chapter four: Dynamics

develop a new gas turbine engine. Compressing the time for new product development has many interesting facets, and we will rehearse some of these later in the book, but one of them appears to be the emergence of "crisis heroism." What we mean by this is the phenomenon of someone or some elite group of people cutting a swathe through corporate bureaucracy and conventional practice in order to get the job done, when time is of the essence and typically the project is massively behind schedule. Such heroes are often known as firefighters. But it came to our attention that while Rolls-Royce has some of the finest firefighters in the world, it also has some of the world's most able arsonists! How so? Because the company breeds a culture of firefighting and crisis heroism, and what better way of propelling yourself to the top as a firefighter than by starting the fire in the first place? Now arson is a pretty strong word, but fires can start simply by neglecting precautions to make sure they do not start, or if they do, that they are quickly extinguished without heroics being necessary. In this instance the burden of dealing with the complex problem of time compression for new product development has been shifted to the simpler situation of a quick fix (see Figure 4.4).

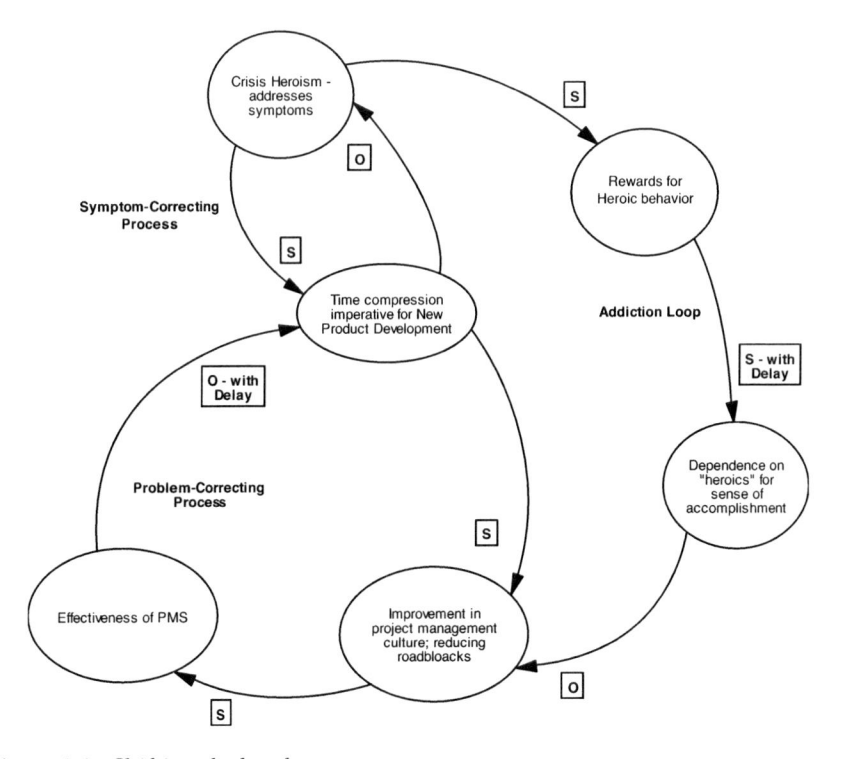

Figure 4.4 Shifting the burden.

In the shifting the burden archetype, two processes are at work: the first is the symptom correcting process and the other, which goes largely unnoticed or unattended, is the problem correcting process. The former is a reinforcing loop, whereas the latter is the proper balancing loop. But because of the thornier issues involved, there is a delay in seeing the effects. Delays, of course, are obviously unwanted by a time compression effort, except that it is an ironic reality that in order to save time on a project, time must be expended or invested in the resolution of the time compression problem. These processes are augmented by additional loops that in effect make the quick fix addictive, obscuring the need or the value of the proper balancing loop and exalting the value of quick fixes. Shifting the burden back to the problem correcting loop might very well be served by the mere advocacy of this archetype.

Our fourth archetype is known as tragedy of the commons, and one of us has personal experience with this situation from a consulting assignment on a project to build a communications payload for Inmarsat. The technology developer of the payload, at the time Matra Marconi Space, had spent years of R&D effort in phased array antennas in anticipation of the potential commercial gain for a giant Telco such as Inmarsat. Then came payday, or maybe payload day. The spacecraft had to be built by GE Astrospace, now part of Lockheed Martin, and Inmarsat had a contract that enabled it to do just about anything to its suppliers. What we are talking about here is a group of the world's best technologists in this arena—there is nowhere else to go—and if these boys can't develop it, it just isn't possible. Subsystem functionality is what technologists love to provide, and with a customer like Inmarsat this is honey for the bread. But Inmarsat only makes money from the operational satellite doing its thing in space, not in the lab or on the shop floor. What is more, launch vehicles are in huge demand, and getting an appointment with one of these for your spacecraft requires long-range scheduling; thus, you do not want to be late for your date or the rocket will carry someone else's load. Launch defines an immovable end stop.

There is more. Penalties for extra weight are punitive, and you can bet that the marketing guy never liaised with the engineering guy when he cut a deal with the customer on mass. So that is the mix: a date in the diary not to be missed, a nonnegotiable payload weight, and a drive toward greater functionality and subsystem performance. It is a toxic potion, and it boiled over at one meeting one of us attended, a preliminary design review for the payload. The chief systems engineer, an irresistibly charismatic and totally uncompromising character, announced to his engineering team: "You're not getting any more mass, any more power, and any more time" ("Do you get me sweethearts?"—our line, a favorite from *As Good as It Gets*).

The chief was making perfectly clear that what each engineer had was access to a common resource—mass, power, and other budgets—and this resource was finite, limited, and would not be expanded. The desire of each subsystem designer was to draw on this resource as far as possible,

Chapter four: Dynamics 73

rightfully so, in order to optimize the functionality and performance of his and her part of the payload. It is natural that each has the attitude that whatever is left, having taken his or her own share, was the other guy's problem. In pursuing something that was good for the customer, they were in fact pursuing something that was bad for one another, and in the end bad for the customer. The tragedy is that this local optimization can cause overloading of the common resource (the commons) and a meltdown of the entire effort. This is explained with reference to the systemic structure shown in Figure 4.5.

This schematic shows two (though there could be more) linked limits to growth archetypes sharing a common constraint or finite limit. In our case the surface acoustic wave (SAW) filter designer is looking for more power in order to increase bandwidth and reduce insertion loss. Likewise, the designer of the frequency-controlled active phased array needs more power in order to optimize beam forming. She wants what he wants; both have reasonable desires and yet their total demands, over time, limit the individual gains each can make. Moreover, this totality further limits the available power budget that then balances any gains each might want to make in pursuing local optimization.

Tragedy of the commons is the system dynamicists' version of the classical trade-off studies highly familiar to systems engineers. What our systems language is doing, however, is pointing us to a coin on one side of which we find the technology, represented, for example, by a limited power budget, and on the other side lies the agency (the designers in our case) whose reconciliation can only be met by a shared understanding of the dynamics of their respective

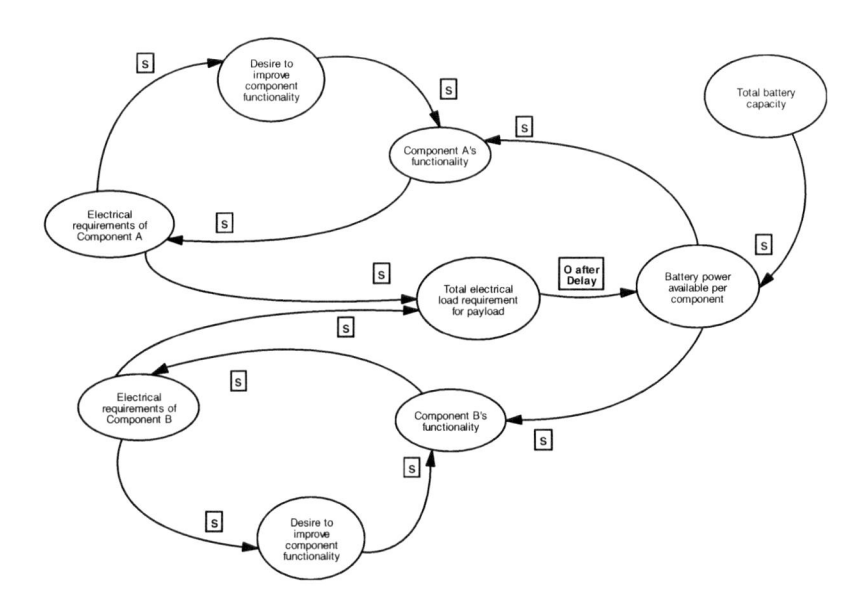

Figure 4.5 Tragedy of the commons.

design pursuits. Increasingly we will find this coin pop up as we look at the dyadic nature of technology realization and enterprise integration.

Our final structure is one that sheds even more light on the lack of cooperation between people who ought to be in partnership, a lack that is accentuated through unintended and often undetected consequences. It is termed the accidental adversaries archetype. To illustrate we will borrow the example for *The Fifth Discipline Fieldbook* by Senge and others.

Proctor and Gamble (P&G) and Wal-Mart are the largest consumer product and retailing companies in the world, and of course do business together. A shared goal would be to improve the effectiveness and profitability of their production/distribution system. In reality this system does not exist, or at least it is not a system that either owns, and in that sense neither owns it, which is why it might be said that it has no reality. Therein lies a potential hazard.

P&G and Wal-Mart have benefited from a working relationship over a long period of time, a feature pointed to by the outer loop of Figure 4.6, which forms a gently reinforcing loop.

In order to boost market share and hence increase profits, P&G, like many other manufacturers, had learned the value of heavily discounting the price of its goods, for which it used lots of price promotions in marketing campaigns. This is shown in P&G's balancing loop. But price promotions create extra costs and difficulties for distributors (like Wal-Mart), and a coping mechanisms is to "stock up," also known as forward buying—buying large quantities of the product during the discount period, selling it at regular price when the

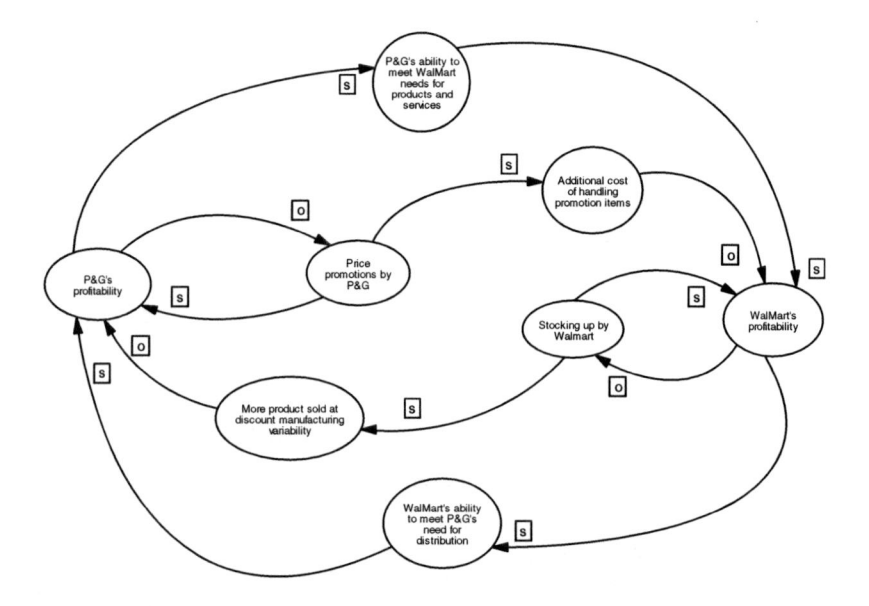

Figure 4.6 P&G and Wal-Mart.

Chapter four: Dynamics 75

promotion ends, and using that extra income to improve margins. (This strategy is shown in Wal-Mart's balancing loop at the lower right of Figure 4.6.) This unhappily undermines the manufacturer's profitability, because the retailer discounted many times the manufacturer's amount of product. Moreover, it leads to wide variations in manufacturing volume, adding to costs, since distributors do not need to buy what they have stocked, and so orders are nonuniform from the manufacturer's point of view.

To compensate, the manufacturers continue their discounting policy, and the retailers their stocking policy, causing a death spiral of mutually detrimental actions. What are the lessons from this phenomenon? Recognize that: All actions have potentially unintended consequences; although it may be "conspiracy," it is more likely unintentional or accidental adversarialism. Each party is inevitably pressed into its own line of (local) optimization, which can (massively) suboptimize the greater good, and hence the collective good. The collective good can only be obtained by developing underlying structures, an objective reality, which each subject can freely inspect and rationalize, thereby laying a basis for collective repair.

4.2 Time to Think

1. The UK government introduced its ban on smoking in public places, for example, pubs and restaurants, on July 1, 2007. Offices and even company cars, if more than one person uses them, are now, by law, designated no smoking areas. England's smokers are following a well-trodden path that has led from glamour through toleration and suspicion to a final destination of pariah status.[4] Whereas it was once thought pointless to tell smokers to quit, since their addiction would clearly prevent such warnings ever being heeded, the dramatic turnaround in the smoking population, down from two-thirds of British men in the 1950s to less than a quarter today, is an incentive to HMG[5] to be even more interventionist. Public places, including churches, are required by law to display prominent no smoking signs or face penalties; sign miscreants may have to fork out as much as two thousand versus four hundred dollars for a rebel smoker. Government stridency goes hand in hand with heavy tax regimes on smokers (seventy-five of the price of a pack of twenty cigarettes goes to HMG) and concern that a national health service free to patients is being clogged up by ill-health that is self-inflicted by foolish smokers. Using this and any other supplemental information you consider useful, create a causal loop diagram that reveals conflicting interest among relevant stakeholders. Analyze this to provide a more enlightened basis for policy making, including judicious taxation regimes, privatized health, and elimination of social inequities.
2. In an effort to ease periods of energy shortages, Americans since the mid-1980s have imported more and more barrels of oil to ensure their daily "fix." Unwilling as a country to restrict use of autos and other

76   *Systems Thinking: Coping with 21st Century Problems*

luxuries, we have grown addicted to foreign oil supplies. The U.S. government has even engaged in a military buildup in the Middle East to secure this long-term source of oil.

At the same time, American scientists have tried to develop options for alternate energy sources. Switching from an oil-based economy to one based on multiple sources poses a challenge. It is difficult to focus on developing alternative solutions when every day the country hungers for more and more oil. As more attention is turned toward foreign oil for short-term satisfaction, less is invested in developing alternative energy sources. Draw a causal loop diagram that shows how energy shortages, oil imports, and developing alternative energy sources influence each other. What does this tell you?

3. Figure 4.7 shows a causal loop diagram indicating plausible relationships between key variables relative to the problem of illegal immigration in the United States. Translate this diagram into a set of statements that have been used to capture what the modelers regard as the principal topics involved. Consequently, what, if any, are the counterintuitive issues that emerge? What does this mean for policy making and prioritization for executive action? What, in your view, are further key variables that this model neglects, and how do these influence the

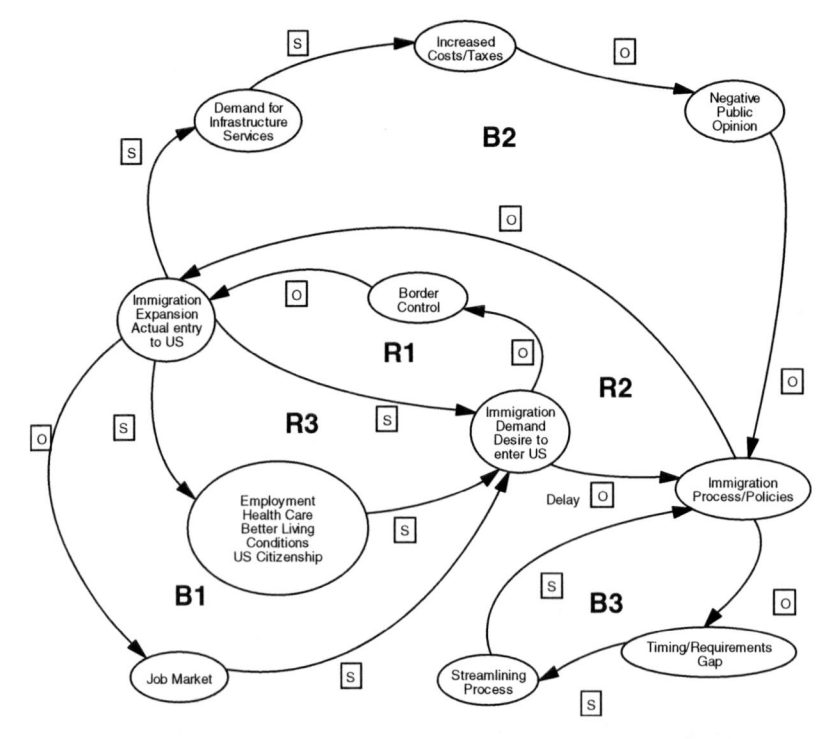

Figure 4.7 Exercise in casual loop.

Chapter four: Dynamics 77

construction of a more relevant model? Construct the model in Figure 4.7 using the rules of causal loop diagramming.
4. The following extract is taken from the executive summary of the report of the Senate Committee on Homeland Security and Government Affairs entitled "Hurricane Katrina: A Nation Still Unprepared":[6]

> Hurricane Katrina was an extraordinary act of nature that spawned a human tragedy. It was the most destructive natural disaster in American history, laying waste to 90,000 square miles of land, an area the size of the United Kingdom. In Mississippi, the storm surge obliterated coastal communities and left thousands destitute. New Orleans was overwhelmed by flooding. All told, more than 1,500 people died. Along the Gulf Coast, tens of thousands suffered without basic essentials for almost a week. But the suffering that continued in the days and weeks after the storm passed did not happen in a vacuum; instead, it continued longer than it should have because of—and was in some cases exacerbated by—the failure of government at all levels to plan, prepare for, and respond aggressively to the storm. These failures were not just conspicuous; they were pervasive. Among the many factors that contributed to these failures, the Committee found that there were four overarching ones: 1) long-term warnings went unheeded and government officials neglected their duties to prepare for a forewarned catastrophe; 2) government officials took insufficient actions or made poor decisions in the days immediately before and after landfall; 3) systems on which officials relied on to support their response efforts failed, and 4) government officials at all levels failed to provide effective leadership. These individual failures, moreover, occurred against a backdrop of failure, over time, to develop the capacity for a coordinated, national response to a truly catastrophic event, whether caused by nature or man-made. The results were tragic loss of life and human suffering on a massive scale, and an undermining of confidence in our governments' ability to plan, prepare for, and respond to national catastrophes.

Use this and any supplemental information you deem appropriate to create a casual loop diagram that models these various issues and further reveals some of the complex interactions that take place in events of this kind and in their subsequent remedy. What can we learn, if anything, from such a model that perhaps our common sense and normal linear modes of thinking do not expressly give us? As simple as this model appears, does it provide testimony to the value of systems thinking to problems of this scale?

Endnotes

1. Senge, P., *The Fifth Discipline*, Currency Doubleday, New York, 1994.
2. Anderson, V., and L. Johnson, *Systems Thinking Basics: From Concepts to Causal Loops*, Pegasus, Waltham, MA, 1997.
3. Ibid., chap. 5.
4. "None So Deaf as Those That Will Not Hear," *The Economist*, June 21, 2007, pp. 62–63.
5. HMG is Her Majesty's Government—that which is partly democratically elected and partly Queen Elizabeth's (II) appointment. See, for example, http://www.imdb.com/title/tt0436697/.
6. See http://hsgac.senate.gov/_files/Katrina/ExecSum.pdf.

chapter five

Soft

5.1 Breakfast @ Tiffs 'n' Ease

There they sit. The man and the woman. At a table littered with the remnants of a lengthy breakfast. Two people scarred by the trials of a lengthy marriage. Abandoned scrambled eggs and unwanted wheat toast grow miserably cold. The memories of joy and laughter of intoxicated newlyweds are obscured by long distance and obstructed by present distancing. Replaced by unwanted approaches and abandonment to separate newspapers of antithetical political persuasions.

The woman tortures herself with vivid imaginings of her husband's secret infidelity and evident indulgences. The man contemplates mild satisfaction of his wife's surrogate fantasies. Briefly. But he doesn't really care. He hasn't nearly traveled the roads in his spouse's mental map. But he doesn't care. He lets his mind wander paths his thoughts trace as hidden eyes follow the paper's chase. Together at the wooden table. Barricaded by their personal pulp fiction. Silence reigns as it would at a funeral.

"It says here," cajoles the wife, "that men who don't drink, don't smoke, and don't chase after loose women live longer."

Another nail in the coffin? A call for the dead to rise? A warning to the wicked? He doesn't care. But he will answer. From behind the fence line of his preferred editorial.

"It serves them right!"

The story is supposed to make you laugh. Without preparing you for humor. Laughter often accompanies the unexpected twist. Something about colliding worlds releasing energy. You are led to believe one thing by getting drawn into one world. Then you are smacked between the eyes by the totally unexpected. An opposite world. It is a device. An invention. Invaluable for systems thinking, whose pivot is simultaneously tenable viewpoints.

We believe, as do many others, that there is such a thing as systems thinking. This is not just a thinking about systems—that would be enough, but, more interestingly for us, it is a thinking that is based on the notions or concepts that essentially define systems as phenomena. It is thinking *from* (or with) systems. The former has systems on the outside, the object of our thinking. The latter has them as a wellspring of fresh thinking, of opposites and paradoxes, of simultaneously tenable viewpoints, a thinking that is "out of the box."

We will present two views of systems thinking, not without risk, by using journey as a metaphor for mind travel, and story as a device to chart our journey. Systems thinking is to the systems journey what seeing is on a literal journey. If you know where you are by correctly interpreting what you see, using electronic means to cover all the senses, you are more likely to make the journey you intend and have richer experiences en route. The problem is that there are many ways of looking at the same thing, and sometimes we are expected to make greater effort to look beneath the surface of what we see, or put another way, to delve more deeply and respectfully into the evidence before our eyes.

Having presented our systems concepts and, in particular, the words that express these in Chapter 2, we now propose to use them, grammatically and in other ways, to put together devices to propel our systems thinking—devices such as techniques for applying our thinking and tools to shape both our understanding of systems and our ability to do systems thinking. Our principal guides here are two Peters: Senge and Checkland. The former provided us with a systems language for seeing, summarized in Chapter 4; the latter, a methodology for seeing with greater acuity and respect, for others and the problems they feel. That is the subject matter of this chapter.

Throughout the book we will continue to amplify systems practice with systems thinking, drawing particularly on complexity theory and the burgeoning science of networks. But we will also recognize the acknowledged giants of both contemporary systems thinking and our heritage of systems engineering. We will congratulate these pioneers. Some are gone, but their legacy endures. And they are succeeded by today's thinkers and practitioners. We are genuinely excited by these efforts and contributions, which we believe will stand the test of time. But for now, we concentrate on simple efficiency.

5.2 Softly, as I Lead You

If there is one vital contribution (in fact, there are many) that Peter Senge and his coworkers have made to the analysis of complex systems, as exemplified in industry dynamics, it is to lay bear the inherent and inevitable tension that exists between perspectives. At one level this is the tension between short- and long-term goals and their accompanying actions—what we might call temporal tension. This is revealed by and in the system archetypes of shifting the burden and fixes that fail. At another level, this tension is seen between individual good and collective good, so beautifully captured by tragedy of the commons and accidental adversaries. This is what we might term contextual tension.

A second Peter, Peter Checkland, will show us a much greater degree of complexity in tension between perspectives, a tapestry of tension that is

Chapter five: Soft 81

revealed among all who have a view on a given situation, be this expressed as a need, a requirement, a constraint, a candidate (solution), or whatever. This group of viewers, known as stakeholders, commonly exhibit tension on a rampant scale, making analysis less amenable and resolution of conflict, even to the point of defining the problem, elusive. Tension among this group is incredibly rife. Nevertheless, quite apart from lamenting the existence of this stakeholder tension, or being frustrated by its manifest potency, we can leverage off this, using appropriate methodological skills, much as the existence of system archetypes offers a means of repair, not just a diagnosis of despair.

A schematic of Checkland's methodology is shown in Figure 5.1. We take the opportunity now to witness this stakeholder tension firsthand and turn it to our advantage in terms of both problem definition and synthesizing culturally feasible change that addresses the defined problem.

5.2.1 *What Seems to Be the Problem?*

Did you ever visit the doctor felling really unwell? Headache, nausea, aches and pains all over your body. Not eating properly, and when you try to eat it all ..., well never mind! You have no idea what is wrong with you except that death would be a relief, but since you might get better and doctors are supposed to help, off you go. Waiting your turn, seemingly interminably while your condition steadily and dramatically worsens to the point where not even death looks like the answer, you finally get to see the physician.

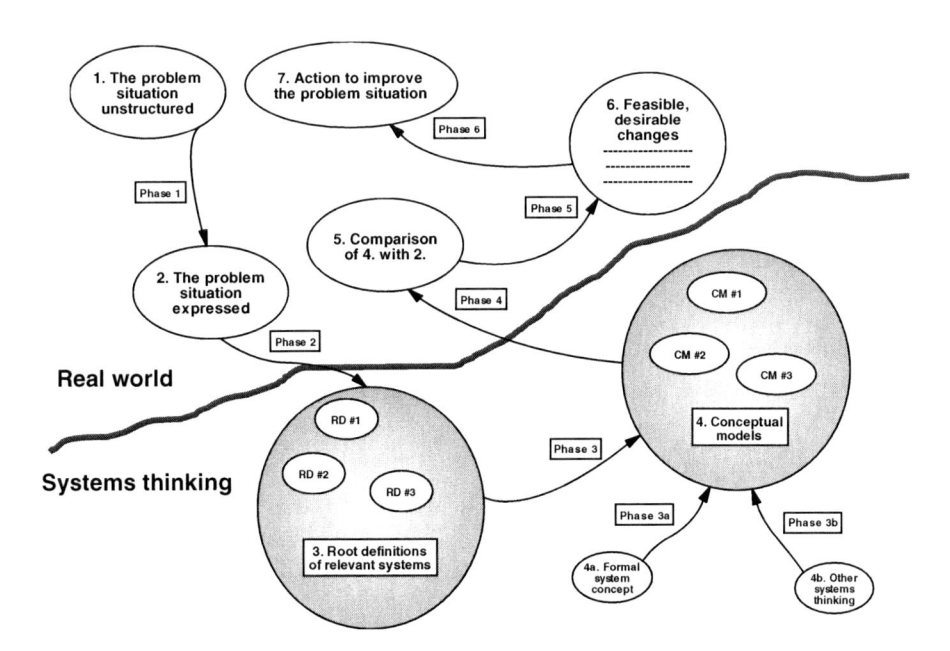

Figure 5.1 Checkland's Soft Systems Methodology (SSM).

And what does she say? "What seems to be the problem?" This is a lady who has spent years and years getting qualified, studying endless texts, passing innumerable tests, and having had the benefit of the best education possible. You feel like you want to die, but you hope that will not happen. You hope she has the answer. But instead *she asks you* a dumb question. What does this mean? Well, apart from the fact that you are going to have to wait a tad longer to be healed, it means that she wants you to tell her how you feel. Where is the pain? What is it like? How long has it been like that? And so on. Symptoms are what they are called, and this lady has the expertise to interpret those symptoms and hypothesize a cause. In a sense she has engaged in a problem of problem definition and is using a rather clumsy opening line to enlist your help, before your demise, in making that process work.

Engineers are natural problem solvers, par excellence. They love to solve problems and nobody is better than them at inventing or conceiving solutions. But what stymies them is the lack of a problem. Clearly, if they do not know what the problem is, they cannot start work. Except that Peter Checkland saw that in what the engineer, and particularly the system engineer, had achieved was to develop a process of problem solving that could be applied upstream, where the outcome would be a defined problem—a diagnosis—and the input to that process would be, for want of a better description, symptoms. Except the symptoms would necessarily be volunteered by a whole bunch of "unwell" people, or folks encountering ill-ease, and the diagnosis would have to respect the validity of all these symptoms even though they may appear to be self-contradictory or mutually inconsistent. Dr. Checkland did a great job. Let us take an example.

History changed on October 4, 1957, when the Soviet Union successfully launched Sputnik I. The world's first artificial satellite was about the size of a basketball, weighed only 183 pounds, and took about 98 minutes to orbit the Earth on its elliptical path. That launch ushered in new political, military, technological, and scientific developments. While the Sputnik launch was a single event, it marked the start of the space age and the U.S.-USSR space race.

President Eisenhower became convinced that the satellite itself posed no immediate military threat to the United States. But the achievement did, in the longer term, and the nature of that threat lay in the battle for hearts and minds of undecided nations as to the course of their destiny, via capitalism or communism. It fell to President Kennedy to make a response.

In July 1958, Congress passed what was commonly called the Space Act, which created NASA as of October 1, 1958, from the National Advisory Committee for Aeronautics (NACA) and other government agencies. The old general had fashioned the tool, but the charismatic patrol boat (PT) boat commander had to put it to work. Imagine the debate as to what the problem was. No shortage of definitions. But with little or no agreement. What was the problem? And how can you define it in such a way that solving that problem would really give the answer to the questions of the age. Which is better—capitalism or communism? Where will space technology lead, in terms

Chapter five: Soft 83

of military superiority? Can the United States' existing military strength be used effectively against a genuine spirit of adventure and exciting technological advance? These are tough questions and the answers hardly straightforward. True, you can throw money at a problem, but where do you throw when you cannot be sure what the problem is? In the end, it was decided that the real challenge was to win the battle for heart and minds the world over. No one could have spoken more eloquently or sharply than the young president from Massachusetts (bold type indicates clues to the definition of the problem):

> If we are to **win the battle** that is now going on around the world **between freedom and tyranny**, the dramatic achievements in space which occurred in recent weeks should have made clear to all of us, as did Sputnik in 1957, **the impact of this adventure on the minds of men everywhere**.... Now it is time to take longer strides; time for this nation to take a clearly leading role in **space achievement**, which in many ways may hold **the key to our future on Earth**.... Space is open to us now; and our eagerness to share its meaning is not governed by the efforts of others. We go into space because **whatever mankind must undertake, free men must fully share**.

He then came up with a solution: to land a man on the moon and return him safely to the Earth before the decade (the 1960s) had ended. His solution then became NASA's problem, and so it goes.

In Checkland's methodology he respects the fact that not all problems come clearly defined, nicely wrapped, and ready for a problem-solving approach. This is not courting paralysis by analysis but rather forestalling considerable nugatory activity. No fixes that fail for him, at least in principle. The initial conditions, if we might call them that, are feelings of ill-ease, poorly articulated, strongly and sincerely felt, and in much need of a "medic" who can intelligently ask "What seems to be the problem?" Of course, then the fun begins because there is no single patient—more like a whole hospital of needy folk. He gives some guidance as to treatment. You certainly need patience (no pun intended) and persistence. You need to respect each and every expression of ill-ease, carefully noting that each is a part of the puzzle; some will be discarded that is true, but not initially and never without that rejected piece having shone some light on the meaning of other retained parts. Since it is a problem that is being articulated, candidate solutions are actually problem definitions and need to be introduced, to the problematique, in a timely fashion and with a keen sense of the political. Knowing the end or the immediate goal, a well-structured problem statement, holds hope that a trajectory can be found from the initial set of symptoms. It is seldom that one can be given such a crisp, clear, and compelling problem statement, humbly accompanied by no remedies. An exceptional example fell to one of us in the guise of an article in *The Sunday Times* (November 1, 1998) penned by Gerald Corbett, at the time CEO of Railtrack, the corporation in the United Kingdom with responsibility for governance of the stations and railroad track used by the public and freight companies. Later on in the book we will cite this as a pristine example of lucidity and political persuasion. It was beautifully written and perfectly balanced in view of the huge diversity of stakeholders and the inherent tensions that existed among them. We regard such statements, from one stakeholder seeking in a genuine spirit of collaboration for the system, as a whole in which all stakeholders are involved. We have seen others and we will cite them fitfully as we make our systems journey.

5.2.2 Getting to the Root of the Problem

At first glance, this might seem like swapping surgeries: from doctor to dentist. Neither of us has had root canal treatment, and we look forward to that being true beyond the sound of the last voice we hear. The dental profession, and in particular the practitioners of endodontic therapy, might vigorously claim that root canal treatment is thoroughly undeserving of its reputation for being a painful process. Nevertheless, we do not want to go there, if we can help it.

Our use of the term *root* here refers to the simpler process, nevertheless painful for some, of creating root definitions of systems predicated on the structured expression of the problem definition emerging from the first phase of the Checkland methodology, known as Soft Systems Methodology (SSM), depicted in Figure 5.1.

The SSM is differentiated from hard systems engineering, the process we amplified in Chapter 3, in the key sense that its objective is to help analysts and stakeholders realize a human activity system that one can associate with a technology development project that requires the SE process. Recall that we said above technology realization and enterprise integration are two sides of the same coin. We reiterate this notion by suggesting SE and SSM are two sides of another coin. As Checkland puts it: "In hard systems analysis the concept is that there is a system to be engineered and this occupies an unequivocal place in a manifest hierarchy of systems." In this sense the engineered system deals with the problem and its purpose is to solve, eliminate, or remove the problem. By contrast, "in 'soft' systems—which include human activity systems considered at a level higher than that of physical operations—there will always be many possible versions of 'the system to be engineered or improved' and system boundaries and objectives may well be impossible to define."

For this reason, the first phase of the SSM is deliberately intended to elicit the richest possible picture of the situation being studied. In the second phase, root definitions of relevant systems, we are dealing with this question: What are the names of notional systems that, judging from the output

Chapter five: Soft

of the first phase—a well-expressed definition of the problem—seem relevant to the problem?

In SSM, natural language is key, in order to express the problem and then begin to nominate candidate solutions. Engineers may be a little uncomfortable with this since typically they prefer calculus, software, or drawings to express themselves. But this cannot be avoided, and gaining confidence in expression through language is not an unobtainable goal. For example, does what Checkland says here make sense?

> Root definitions thus have the status of hypotheses concerning the eventual improvement of the problem situation by means of implemented changes which seem to both systems analyst and problem owners to be likely to be both "feasible and desirable." To propose a particular definition is to assert that, in the view of the analyst, taking *this* to be a relevant system, making a model of this conceptual system, and comparing it with present realities is likely to lead to illumination of the problems and hence to their solution or alleviation.

We think so. And our illustrations will hopefully help.

A huge concern in the United Kingdom is over the rise in crime and the treatment of convicted criminals. Some people argue that crime will increase if the law does not deal severely, but justly, with the guilty. Evidence exists to show that the majority of crime is committed by felons, who live a life of crime and cannot break free of its grip. If these folk could be persuaded in some way that crime does not pay, then their activities would come to an end, crime would fall, and the low level of crime would naturally keep the lid on its increase. High crime breeds a criminal mentality, whereas low crime keeps people mostly honest. Some would say. So we ask, at this point, given that general expression of a problem, what are feasible and desirable root definitions (for future hypothetical human activity systems) that we can conceive, relevant to this expression?

A crucial piece of the justice jigsaw is jail. We have all seen *The Shawshank Redemption* and the injustice done to Andy Dufresne, by both the courts and the prison regime, is enough to keep most of us honest all of the time. If a jail should not be like Shawshank, what should it be like—in the interests of reducing crime? What is a jail anyway? Some people say that it is a place to administer correction, so that those who go will, when they leave, never want to reenter. Others say that it is a place of incarceration, temporarily or permanently, from which there is no escape, ensuring that law-abiding citizens are not menaced by escaped felons. Another group may argue that it is a place of rehabilitation where people can learn the errors of their ways, discover the flaws in their character that can then be repaired, and find healing for their life from the servants who administer the prison, so that felons will never again turn to crime. A radical group may argue that a jail is a university of crime, a place where convicted felons can learn more about the profession of crime, from leading experts, who for no reason of their own have been unable to elude detection and arrest. If that were so, it might be worth breaking into jail in order to acquire the expertise to make it big when you land back on the outside, bigger than you would without the benefit of such a university education, carefully adding to your knowledge that which will help keep you out a second time. Now while these perspectives may not be entirely incompatible, they do make unpleasant bedfellows. Pity the architect who has to come up with a jail, a real physical system with its associated human activity system, that realizes all these tenable viewpoints simultaneously. That is why Checkland offers a methodology, which includes root definitions, but has more. Before you move on to the more of SSM, take the following root definition as an example of what a social services department might be, and having savored it, attempt to come up with something equally appetizing for a jail:

> A department to employ social workers and associated staff to build and maintain residential and other treatment facilities and to control and develop the use of these resources so that the social and physical needs of the deprived sections of the community that government statute determines or allows, to the extent to which local government, as guided by its professional advisers, decides is appropriate, are met within the annual capital and revenue constraints imposed by the government.

5.2.3 Ideally, This Is What We See

The next phase of the SSM turns root definitions into conceptual models, as shown in Figure 5.2. This is the part that engineers will most enjoy since it is a building operation.

Yet again, however, it is building with words; but for system engineers who over the past 10 years have gone to town on requirements elicitation and their management, to the point where the field is actually now called requirements engineering, words are not nearly the problem they once were. The operation required in this phase is to turn definitions that express *being* into models that capture *doing*—an emphasis on activity that is required, expressing a social system in action. The conceptual model, pertinent to a root definition, is an account of the activities that the system must *do* in order to be the system named. The structural elements for the model are derived from the root definition; the dynamic character of the model is facilitated by the formal system concept, expressed in diagrammatic form below and by other systems thinking. Figure 5.3 is a representation of this line of thinking.

In creating conceptual models there are some key pointers to keep in mind. First, the models are not intended to be models of reality. We are not trying to capture an "as is" in the classical sense. For one reason the "as is"

Chapter five: Soft

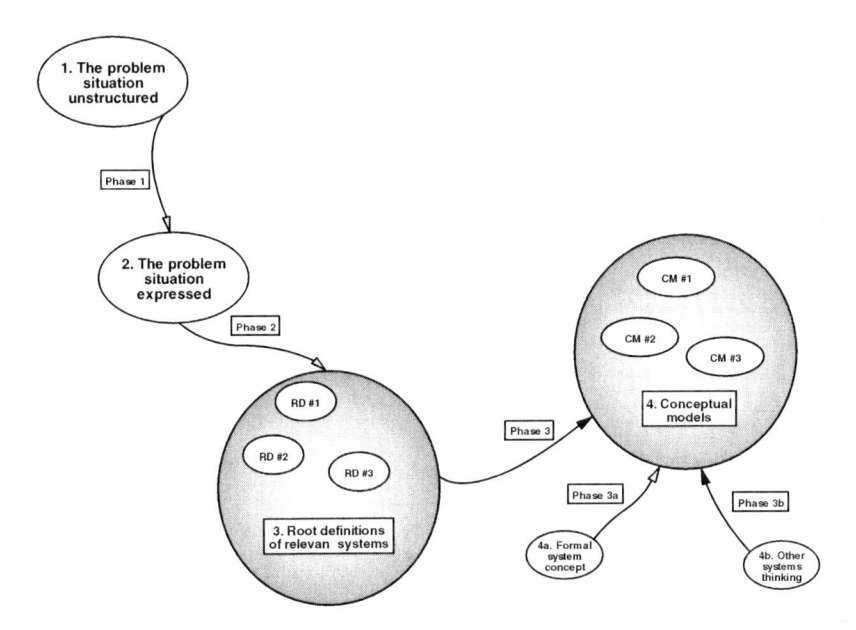

Figure 5.2 Checkland's phase 3.

may not exist, and anyway part of the problem with an "as is" that does exist is that it should not continue to be; it is the primary focus of change. Checkland argues that the best way to make this happen is to divorce reality from systems thinking and to create models of *ideality* that can be insightfully compared with a known situation. This can be difficult for hard-nosed engineers, but we argue that it is liberating and it certainly taxes to the limit the engineering domain expertise that is needful for drawing upon to construct models. For this reason, nothing ought to be included in the model that cannot be justified by reference to the root definition. This is why phase 2 is so crucial; it underpins the success of this phase. The primary elements of the model are verbs. The technique therefore is to assemble a minimum list of verbs covering the activities that are necessary in a system defined in the root definition, and to structure the verbs in a sequence according to logic.

Later in the book we will describe our own original contribution to this technique and show how the SSM itself is adapted by this unique form of representation to tackle the challenges that systems engineers currently face, such as system of systems and extended enterprises. An example of such a conceptual model (which we term *systemigram*) is shown in Figure 5.4. This diagram is based upon root definitions extracted from the article by Gerald Corbett that appeared in *The Sunday Times*, in which he attempted to draw together the elements of a complex rail passenger system in the United Kingdom so that it could better provide the service for which it was intended. Notice that the "bubbles" in this diagram are nouns and that the links

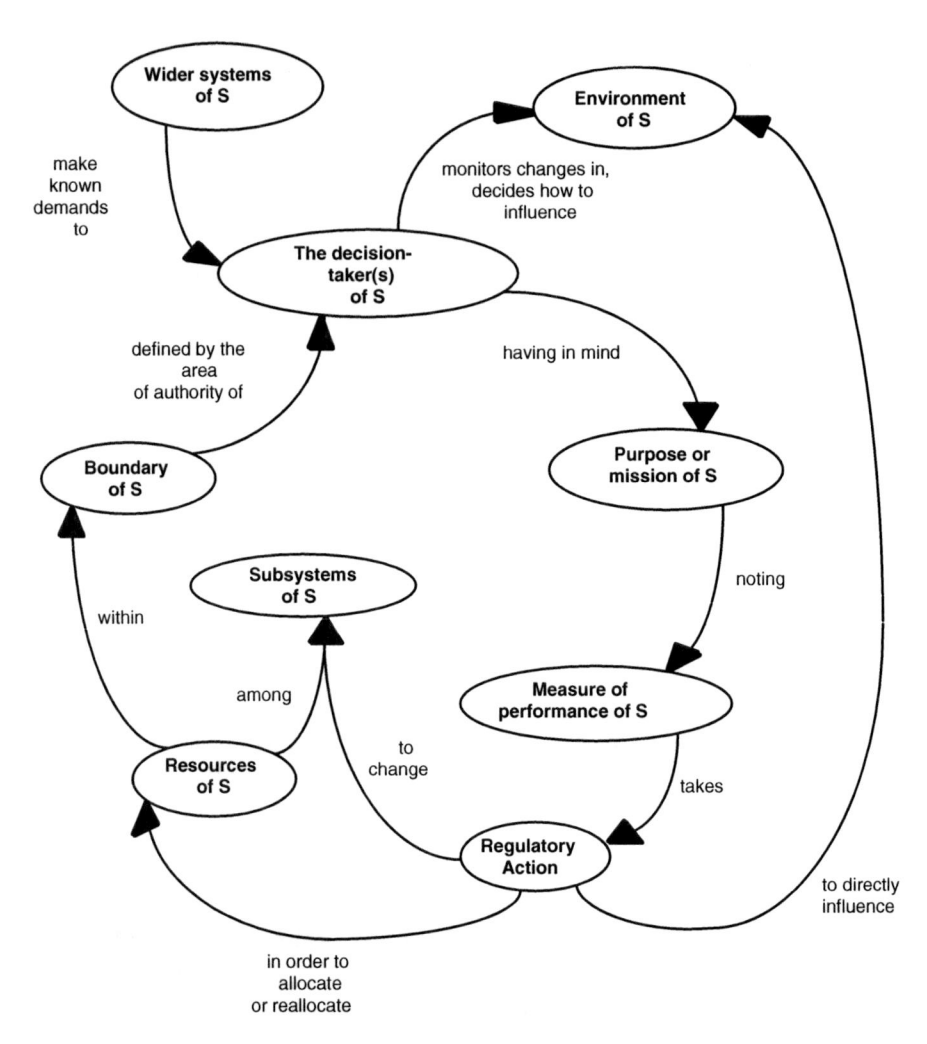

Figure 5.3 Formal system model: Peter Checkland. (From Checkland, P., *Systems Thinking, Systems Practice*, Wiley, Hoboken, NJ, 1999. Reprinted with permission of John Wiley & Sons.)

connecting them are the verbs; so the integration of the system elements is achieved by concentrating on the activities that need to be present in order for the system to do that for which it exists.

Likewise we will later provide our own ideas of other systems thinking that we have found most useful in building conceptual models and in adopting a Soft Systems Methodology.

By inspecting the systemigram above, and by examination of the social services department root definition, we invite you to create your own conceptual

Chapter five: Soft 89

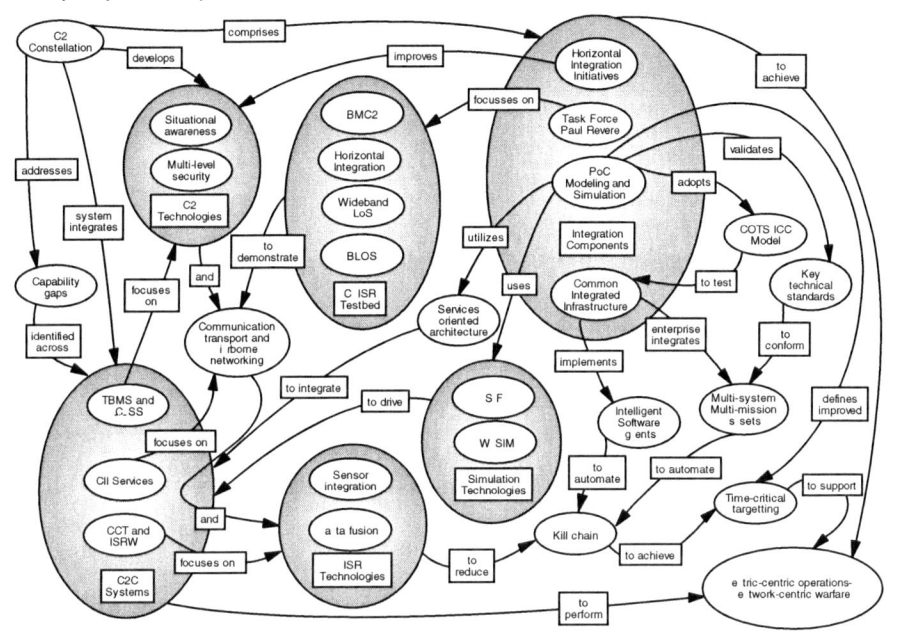

Figure 5.4 UK Rail Systemigram.

model of a human activity system that can do the things it must in order to be such a service.

5.2.4 It's Good to Talk

According to the next phase of the SSM it is time to reenter the real world. The importance of the great divide between the two is to enhance the development of genuinely systemic models that can benefit from domain knowledge yet be uncontaminated by the very things that plague the mechanisms that operate for real people in real situations. In this way it is hoped to shed real insight onto the reasons for the failures, some of which may very well have been caused by "fixes" and the unintended consequences of remedial action. In a sense it is like draining the swamp without the threat of alligators distracting the operation. Technically it is called action research, in which the research takes place by involvement in the real world, as opposed to the laboratory or test bed, and the researcher gets affected by the action. The observer is part of the observation and the observation is influenced by the observer. Recognizing this interplay is important; the separation of the systems thinking world from the reality it addresses actually helps the interplay.

Once again there is a key pointer to bear in mind as the dialogue unfolds among analysts and stakeholders, a dialogue enriched by the conceptual models. This refers to the *manner of making comparisons* between the original problem expression and the conceptual models, and Checkland suggests four distinct ways. First, the models can be used to suggest a line of ordered questioning. Quite possibly what takes place in the model in no way resembles what presently transpires, but that can be a good thing, and it stimulates fresh thinking by the stakeholders. A second way is to reconstruct the past and compare history with what would have happened had the conceptual models been followed faithfully. In both these cases, the models themselves can be hidden from the stakeholders, so that they are not seized upon as either the answer to their problems or absurd notions that can be flatly rejected bringing an intransigence among stakeholders and a resignation to cope with what they have. A third way is to reveal the models and accompany their presentation with questions about how they differ so much from present reality and why. Finally, an approach known as model overlay can be tried. Here a new set of conceptual models are created. This time they are based on reality and are designed to capture as much as possible the way things are. The only rule is that so far as possible they should have the same form as the "divorced" conceptual models. What this overlay approach does is to highlight the distinctions, which of course are the source of discussion for change. What is more, these new models can be reverse engineered into root definitions, and then they can be compared with the one that was obtained from phase 1.

The purpose of this dialogue phase is to generate debate about possible changes that might be made within the perceived problem situation. In practice, the work done so far can itself become the subject of debate and change. This requires humility on the part of the analyst: How can he expect stakeholders to change when he himself will not? Changes can be to any of the artifacts of the process or to the process itself. (But then things can only get better!)

Checkland suggests there are three types of change: in structure, in procedures, and in attitude. Structural changes may occur to organizational grouping, reporting structures, or functional responsibility. Procedural changes are to the manner of getting things done, for example, the periodicity or medium for reporting. Attitudinal change is of the mind and the heart, and usually less easy to accomplish than the former. The criteria for suggesting and effecting change must be whether changes are systemically desirable and culturally feasible. The former refers to a respect for the integrity of the SSM and of all the artifacts this generates; in other words, change ought not be arbitrarily effected simply because "something has been done" but that something has not been acknowledged. The latter criterion shows respect for the problematique and for the stakeholders themselves. Even when change is obvious and agreed upon, it may not actually be deemed implementable simply because of conditions. That is political (and economic) reality. The analyst has not necessarily failed; after all, he is no longer dealing with buttons in the engineering sense, but belly buttons.

5.2.5   The Long Unwinding Road Map

This is not merely a postlude to the SSM. It is in the nature of soft systems that changes produce unforeseen (and unintended) consequences. Even when such matters have been taken on board a priori. Social systems are notoriously nonlinear, which is one reason for introducing complexity theory into systems thinking. More anon. What is even more interesting is that the implementation of change may produce a problematique that is susceptible to SSM. At this point, we hear many engineers cry "paralysis by analysis," and clients complain: "Typical! Consultants!" But the decision for continuation rests with the stakeholders, and that will be influenced by the quality of what has been done and the growing respect for the fact that what needs to get done is not obvious, nor is determining it trivial.

What SSM does is to break analysis and synthesis away from the strongholds of quick fixes, short-termisms, simple mindedness, and singular action. It pays respect to complexity, variety, perspectives, nonlinearity, stakeholders, and counterintuitiveness. It has humility for its origins, being largely borrowed from hard systems analysis and engineering, and is open to usage—the phases can be conducted in any logical order and started at any reasonable point. Methodology is not method.

Whereas Senge's systems language is offered to help achieve efficiency, doing things right, Checkland's SSM is an aid to effectiveness, doing the right things. Both are needed and both can use each other. The question remains: What is right to do? That, for some, is a matter of ethics. On Pilgrim's journey he met people who knew what was the right thing to do, but they did not agree. Can SSM be used to determine what is right, above the level of what are the right things to do? Does SSM have something to say about ethics? Or do we need to ascend to an even higher level?

5.3   Time to Think

1. A middle-aged lady from England is on vacation making a tour of U.S. cities, among them Orlando, Florida, and Atlanta, Georgia. In the latter city she engages in conversation with a hard-nosed engineer, a teacher at Georgia Tech and a native of Utah. They compare lifestyles, though she more vocally than he, the strong and silent type and somewhat cynical even though he has traveled Europe and Asia extensively, while this is her first visit to the United States. She remarks at the end of a long eulogy of her native land, "There's no place like England." He replies, "You're right about that!" What worldviews are being shared here?
2. A man sleeps soundly in his bed next to his beloved wife. Their three children are all safely tucked up in their beds. A family at rest. Suddenly three men each carrying machine guns burst into the bedroom and rouse the man from his slumbers. One of them announces that the government of the country, headed by a ruthless tyrant, has

been overthrown. They demand to know whether the man was loyal to the tyrant or is now in favor of the incoming regime. His life depends on the answer he gives. He declares his loyalty to the new rulers, rejoicing at the fact that the former government has been overthrown and the murderous tyrant deposed. The leader of the gang of three shoots the man dead. They tell his widow: "We always suspected him of being disloyal." There has been no regime change. The bereaved had better learn a lesson. What is the problem here? Express it as richly as you are able in fewer than two thousand words.

3. Imagine that President Kennedy's administration came up with a problem definition as follows:

> We have had less than 20 years peace since World War II ended. Since then, we have seen the rise of a superpower in the east, the Soviet Union, which enslaves vast areas of Europe under a tyranny that is opposed to all the freedoms we hold dear in the West. We are witnessing the emergence of a new world, in Africa, the Far East, and South America, in which peoples have a fundamental choice to make between freedom and prosperity, and a communist regime that we know will trap them in a poverty that begun with our own brand of commercialism. And as these peoples look for signs as to which directions to take and make their destiny, they see the might of the United States grounded while the Soviet Union send men into space, giving them a vantage point from which to proclaim superiority not only for themselves, but also for their system of governance—an announcement that we cannot ignore. Our nation has been built on a pioneering spirit, a love of liberty, and an inventiveness that has tamed the land and established markets to bring prosperity to all. We cannot lose these values or have them held prisoner of outrageous darings by either our potential adversaries or the physical laws of matter. We cannot stand still or be idle while others advance and in their advance make followers of undecided nations. Courage, discovery, and talent made us what we are. It can make us better. And it can persuade others of the justness of our cause and the superiority of our beliefs. Our problem is to win the battle for hearts and minds by bravely and ingeniously pursuing a vision that embodies our free spirit and encourages freedom for all.

Use this text and whatever other background information you can assemble to create hypotheses (root definitions) that will describe the being of notional human activity systems whose realization in time will help ensure that this problem definition is addressed.

Chapter five: Soft

4. The following is extracted from an assignment by a master's student who elected to provide his version of the inaugural speech of January 20, 2009—in the fall of 2006!

> As much as there is to do at home, there is as much to consider around the world. There are two things that we already understand very well. Poverty and disease ravish too much of the world, dealing out harsh conditions to millions and contributing to many of the other problems like terrorism that we see in the world today. President Kennedy nearly fifty years ago said that "man holds in his mortal hands the power to abolish all forms of human poverty and all forms of human life." Since that time, our attention to the abolition of human poverty has fallen short while war and conflict continue at unacceptable levels. Although we must always protect ourselves, we must also shift the balance of the powers that Kennedy spoke of towards abolishing human poverty. The key is to change the framework.
>
> We must break down barriers to building relationships. In some cases, the barrier is in the form of a prism. We stand opposite other countries and each see a distorted image of the other. We must move from behind this prism. If we can do this, America will be part of an axis of understanding.[1] This does not mean, and let me stress this point, that this does not mean a retreat from American ideals. We are a strong nation, capable of defending itself, and willing to bear any burden to secure our freedom. Our military is the strongest in the world, and we owe them an unending debt of gratitude for the work that they have done and continue to do. What it does mean is that we will work to better understand countries and their issues. While we love and cherish our democracy, we also understand that democracy in a country like Iraq can reduce the freedoms of a minority party. Countries must strive to provide more freedoms to their people, but America is mindful that democracy can have chilling effects when elected leaders fail to provide vital services and to protect minority rights.[2]
>
> This balance allows us to move forward as a nation. Without it, the polarization of the parties and lack of consistent action will continue. We can promote democracy on the one hand as a key principle, and on the other hand recognize its limitations in certain circumstances. We can be a leader to the world, and yet let others take the lead on key issues. We can unify this country on basic principles, while disagreeing fiercely but respectfully about others. Our country, our democracy is strong enough to do all of these.

Use this text as a basis to create conceptual models using Checkland's SSM and any other form of system modeling technique you prefer, and compare these idealized models with reality as you perceive it. Use this comparison to create an agenda for change that is both systemically desirable and culturally feasible.

Endnotes

1. The designation "axis of evil" has a chilling effect on diplomacy.
2. Democracy in young or unstable nations often upsets balance of power and threatens minority rights (Iraq is a prime example).

chapter six

Systemigrams

6.1 Into Great Issues

One of us (John Boardman and the *I*, *me*, and *my* in this chapter) was 14 years of age when the youngest elected president of the United States of America rallied the nation with the challenge to land a man on the moon and return him safely, before the decade (of the 1960s) was out.

John Kennedy cast the grand vision and inspired the giant leap, but he did not witness the one small step. John Boardman, however, just before turning twenty-three, watched the president's man, Neil Armstrong, descend the ladder to a new summit of mankind's achievements, on a tiny black-and-white TV set during a summer holiday in the southwest of England.

As the 1960s, widely known for adventures and challenges of all kinds, came to their end, a young generation fueled by discovery and unprecedented excitement wondered what would come next, and how would they play their part, especially since the underlying motivations for President Kennedy's speech, to defeat communism in the cause of freedom, had still to be met.

As I reflect on those years I must confess that at age fourteen calculus limited my vision of integration. I could integrate simple functions of x and determine correct solutions to many mathematical formulations, but did I even know of the integration problems Robert Kennedy struggled with in Mississippi, let alone be able to contribute to their solution? I think not. However, as Armstrong and Aldrin—how characteristic of the United States to field their A team on Apollo 11—walked on the moon's surface, I knew firsthand a little more about the thornier integration problems in life than those of calculus. I had been married for 2 years and we had a 1-year-old son. Families, especially to the immature adult that I was at the time, presented a challenging and ongoing integration problem all of its own. But some problems you live with, and therein lies one solution.

Integration is the word we use to introduce the subject matter of this chapter, which is systemic diagrams (referred to as systemigrams): what they are, how they are created, who would want to use them and why, and where they are headed as a decision-support tool, in our opinion.

In the late 1980s I became involved in an enterprise that presented me with three distinct types of integration problem: political, technical, and conceptual. Twenty years later we are convinced that these types of problem continue to impact us, interdependently, and little wisdom seems to have been received in the interim to distinguish between them, let alone solve them, a meta-integration problem that emphasizes the complexities we confront.

The enterprise I refer to was a European R&D project called ATMO-SPHERE.* This was a project that received relatively huge funding from Brussels as part of the general technology framework known as ESPRIT, which provided its own brand of political and cultural alignment. European leaders were always prepared to look toward political integration but watchful naturally of national sovereignties. An interesting word that was bandied about at that time was *subsidiarity*, something that might not be well known but is expertly practiced in the United States. For this nation, it means the reverse delegation by states for essential federal effort that benefits the United States as a political union. Easier for the United States, a civil war notwithstanding, than for European nations.

So *political integration* is on someone's agenda somewhere, and this in some ways affects the thinking of corporations and their employees, who are also individual citizens of course. These individuals are employees primarily because of their technical expertise and zeal, something that can transcend cultural and political differences. European leaders understood this well enough; so ESPRIT served two goals: a helpful cultural mix creating experiences for dialogue over future political union, and the genesis of a critical mass in IT to begin to match the overpowering might of the United States, embodied at the time in the form of IBM and the lesser giant minicomputer manufacturers (e.g., DEC). Apple Computer was still an infant corporation and the PC largely regarded as embryonic. How things change. Some things.

ATMOSPHERE was seen as a lodestar to guide European efforts toward more powerful computing paradigms and more powerful corporate assets in IT development. Basically, the major IT players in Europe together with smaller-niche firms and some university groups were pooling their expertise in order to build a single software development environment in which tools and application could be built to serve the various needs of the players. A comparable environment was Portable Common Tool Environment (PCTE).[1] Thus, the project itself was a serious attempt to achieve *technology integration*, indeed to provide an environment for the integration of technology systems, both hardware and software. Seen as something of an expert in systems engineering, and therefore a specialist in integration tools and techniques, I was engaged by the EEC as an advisor to the project but reporting to its paymasters also.

Naturally enough I wanted and needed to know more, and ideally before I met the project's management team—so that I would not look too stupid and

---

* The most impressive acronym I have ever encountered! It stands for Advanced Tools and Methods of System Production for Heterogeneous, Extensible and Robust Environments. Its corporate members included Siemens, Philips, Bull, Olivetti, and GEC Marconi. It ran for 3 of its intended 5 years.

Chapter six: Systemigrams

ruin any chances I might otherwise have of bringing some benefit to the project. My introduction to ATMOSPHERE's mission, motivation, and management structure came through reading the proposal on which it was founded—the document that had been submitted to the EEC to secure funding.

I later learned that the project booklet was single authored but that person had left the project before funding had been put in place. His writings were the sole means of communicating his intentions to the project. This failed. Not because the writings were poor, but because they never got read. This is a problem. Communicating strategic intent, especially when it is intelligently written, cannot be entrusted to the writings alone, nor to the presentations of the author. Additional support is needed—support that is faithful to the statements expressing the strategic intent, but value adding in ways that the author points at in his writings. This experience was the genesis of my *conceptual thinking*.

I wanted to encapsulate my understanding of the proposal document in more than an executive summary; it already had one, what would mine add? I also wanted to illustrate that understanding by graphical means, but once again I wanted to be different from the many diagrams that the proposal document contained. The question again arises: How do I achieve this difference, this complementarity? How do I add value? I needed my diagram to be a system in its own right. The whole project was about systems and systems integration. I saw systems everywhere I looked. Not everyone did, or would or could. But I wanted them to see my system—not just a diagram, but a system, one that could point the way to how they could see their own systems, and where they saw nonsystems to be able to repair, redeem, improve, or enhance.

My first step was to capture for myself the essence of the proposal and to do so in words. As I wrote, I was very conscious of the significance of certain words and the significant relationships that these words had to one another, syntactically and semantically. I was working systemically, not just systematically. I highlighted the significant parts (noun phrases) and their significant relationships (prepositional or verb phrases) in two different colors. Diagrams for me are essentially networks having two elements: nodes and links. Some argue there is a third element: text. But my decomposition into parts and relationships was all text; putting them back together in diagrammatic form was another, value-adding way of presenting the text. I did not realize it at the time but what I was doing was creating systemic diagrams, I called them systemigrams, based upon three influences: my exposure to systems thinking, my experiences in systems engineering, and my growing awareness of the complexities of communicating and executing strategic intent.

I decided that diagrams, which could also be regarded as systems in their own right, was the way forward, and that the components of these systemic diagrams (systemigrams) would come directly from the author himself, via his writings—the concepts, constructs, relationships, and emergent features that the language, the grammar, the semantics capture.

So the problem that needs to be addressed when systemigrams come into play is this: a complex project operated by a heterogeneous team searching for a common culture and requiring vision to be articulated and translated when communications between vision and tactics is fraught by lack of ubiquitous leadership but has available a well-composed statement of strategic intent that deserves additional value being added such that the leadership emerges, the involvement of the team is solicited, and the common culture forms by virtue of the ownership of the value-adding proposition.

A major source of inspiration for systemigrams came from a diagram in Peter Checkland's book *Systems Thinking, Systems Practice*,[2] as shown in Figure 6.1. What struck me most about this diagram was that it was both prose

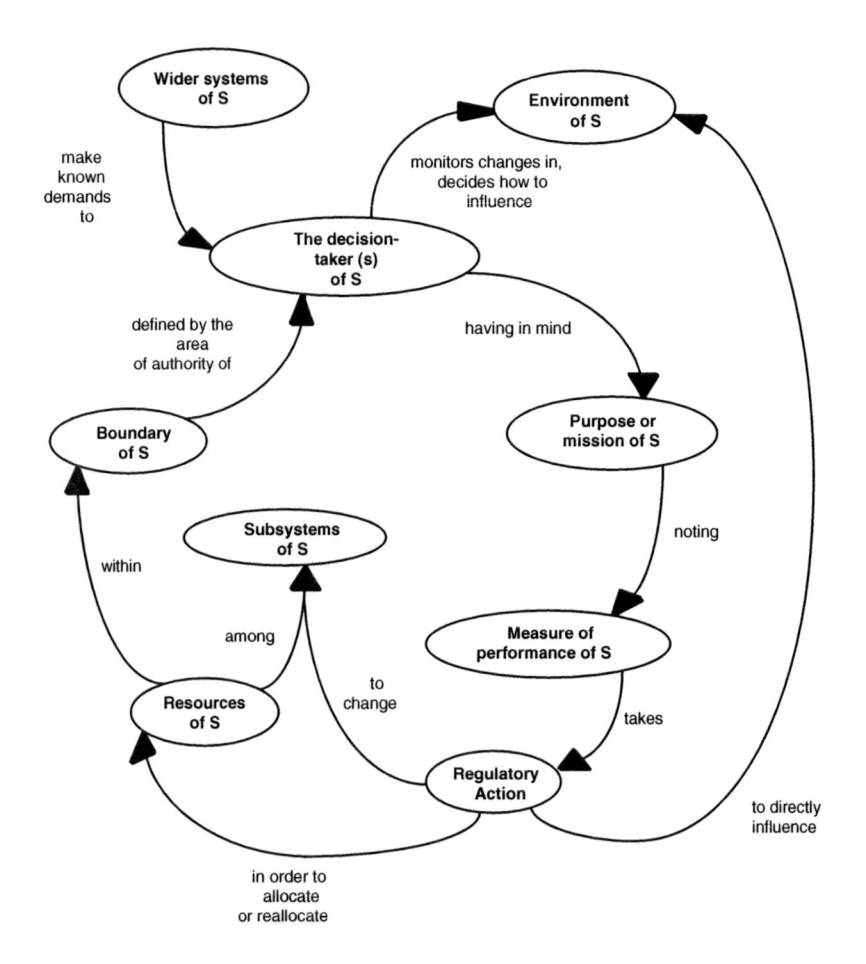

Figure 6.1 Formal system model: Peter Checkland. (From Checkland, P., *Systems Thinking, Systems Practice*, Wiley, Hoboken, NJ, 1999. Reprinted with permission of John Wiley & Sons.)

and diagram, and therefore had the value of each within it. Most diagrams do not read well. You could "read" this diagram. Prose does not convey what a picture can. This picture clearly identified the key elements of a formal system model. I reasoned, if a device could be found, like this diagram that had the best features of prose and graphics, but further was assembled in the spirit of systems—parts, relationships, wholes, emergence, flows, inputs, outputs, transformations, process, networks, and so on—then surely such a device would be a valuable medium and a contribution to systems thinking and systems practice.

It would have to operate at fairly high levels since grammar is important, whereas at lower levels syntax is prime. It would have to be faithful to the text whence it came. It must be possible to "recover" the sense of the original prose by an inspection of the diagram. It should be possible to discover new ideas from the diagram that perhaps a linear reading of the text would not provide, though to intelligent readers digesting great prose, such new ideas would hardly come as a surprise.

I realized I was onto something when it came time for the project's first review, upon its first anniversary of funding. I remember like it was yesterday and could go into graphic detail (no pun intended). In essence the paymaster was livid. He had spent a ton of money and all he had to show for it was a one-page fax explaining a few simple tasks that had been undertaken—or not. He waved this furiously above his head, which could so easily have been adorned with a black cap so far as the project leaders were concerned. He then, in his other hand, waved another single sheet of paper that bore my systemigram, my value-adding comprehension of what ATMOSPHERE was supposed to do, how, and why. He had paid me 0.1% of what he had spent on this ill-begotten mess he was rebuking, yet it held the key to escape his wrath. And so it proved. One measure of success as regards the use of systemigrams to help the project ATMOSPHERE continue to receive funding and produce useful results is worth reporting. When ATMOSPHERE finally closed, its managing director then began to operate a successful consulting business helping other EEC-funded projects to exhibit well at reviews by using systemigrams to communicate and confirm strategic intent! You do not always know the disciples you make.

6.2 Evolution

Of course, diagrams that try to capture concepts are not new, for example, concept maps,[3] fishbone diagrams,[4] Senge's diagrams,[5] influence diagrams,[6] and even the original flowcharts. The one thing about all of these, though, is that they are largely memoryless. They capture the immediacy of prose but then forget that and move on to the next local piece of knowledge. It is more difficult to find longer thought threads in these diagrams since they concentrate on linear thinking rather than holistic thinking. Senge's diagrams are a possible exception to this, but these are always kept deliberately small, and even when these get big it is hard for the reader to make sense of the totality of the language that the diagram conveys.

Systemigrams are based on a complete respect of the totality of prose, believing that its richness deserves to find graphical expression, and in that graphical expression inspire further detailed grammatical exposition leading to more detailed graphical description. The existence of systemigrams as a value-adding proposition, one that will reveal the inner meanings of strategic intent and help build a greater shared understanding in a growing community of people, should force up the ante for defining strategic intent more completely, more thoroughly, more thoughtfully, and more purposefully. The two go hand in hand—excellent prose and great graphics—together supporting the translation of strategy into tactics. Systemigrams are the sine qua non of strategy bridge building.

The progress of systemigrams, over almost 20 years of development, has followed an evolutionary process, involving several Ph.D. students, faculty, and industry champions. Looking back, there appears to be three distinct phases in the evolutionary process:

- Concentration on graphical portrayal of structured prose
- Development of methodologies that use systemigrams for architecting purposes, for example, extended enterprises or business process architectures
- Development of systemigram technique for drilling down from architectural vantage points into detailed consideration of solution implementation

These phases have been roughly chronological with some obvious overlaps, although since there never was a prescient plan to form a critical mass of systemigram developers/users/customers, the evolutionary process has been an uncontrolled one to date. However, the distinct phases have influenced the generation of rules and guidelines for constructing and using systemigrams. In the remainder of this chapter we set out, by principles and practice, the rules for constructing systemigrams, their role in the system architecting process, and how families of systemigrams can provide detailed requirements for systems engineering.

6.3 From Prose to Picture

The top-level requirements we foresee that guide the systemigram construction process are:

- To faithfully interpret the original structured text as a diagram in such a way that with little or no tuition the original author, at the very least, would be able to perceive his or her writings and, additionally, meanings.
- To create a diagram that was a system, or could at least be considered a system in its own right. Thus, if the original structured text could be considered a system, then its faithful interpretation as a new object

should also be systemic, but with features not possible with prose alone, but quite amenable as a graphic (or picture).
- To ensure not only compatibility between the graphic object and structured text, but also synergy so that both objects could evolve into more potent instances, capable of improved dissemination, and community building, development, and mobilization. Thus, it is not a case of either-or but both.

The foregoing imposes rules or conditions upon both objects. First the prose. This must be excessively intelligent. It must be about strategic intent, not procedural tactics. The text is not a checklist but rather a well-crafted piece that searches out the minds of its readers and stretches the mind of the author. Sometimes this text does not exist, but can be brought into existence by facilitation and dialogue with those who own the strategic intent conceptually. Care must be taken, however, that the crafting of the piece is not overly influenced by those with much literary genius but little domain expertise. It need not be lengthy (e.g., two thousand words is quite sufficient); however, it can be a large document—but this would need to become an executive summary of the scope of the systemigram meant to address the scope of the document.

Next for the graphic. This is to be a network, having nodes and links, flow, inputs and outputs, beginning and end. This *must* fit on a single page. Key concepts, noun phrases specifying people, organizations, groups, artifacts, and conditions will be nodes. The relationships between these nodes will be verb phrases (occasionally prepositional phrases) indicating transformation, belonging, and being. Nodes must be unique. Some nodes can contain other nodes, for example, to indicate breakout of a document or an organizational/product/process structure. The network must be legible so that this limits the number of nodes and links. There should be no crossover of links, improving clarity. This constraint further lends itself to systemic design. Such a network tends to be of an interconnected kind for which the ratio of nodes to links is 1.5 or thereabouts. For a systemigram of 20 nodes, the total number of possible links is 190, whereas the actual number will be about 30. This ratio is about 15%, which is held to be the optimal ratio of interfaces in a system relative to how many there could be.[7]

The main thrust of the systemigram, the mainstay, should be diagonal flow from top left to bottom right. This is determined by the chief message of the text. The geography of the systemigram can be exploited to say, for example, the motivation for the strategic intent, its mission, and how it will be accomplished—its management. There will be relationships between the elements of each of these—the why, the what, and the how. Such elemental relationships are invaluable for maintaining coherence for accomplishing the strategic intent. Color can be used to draw attention, in a consistent way, to subfamilies of concepts and transformations. However, the finished systemigram should be aesthetically pleasing and in line with the three top-level requirements, which moderates its form.

If the above description gives the reader confidence to embark on an initial adventure into this space, one that is relatively straightforward, then the first exercise appearing at the end of this chapter can be attempted. The text provided is the product of reverse engineering of an actual systemigram that was constructed from considerably more original text that itself had been culled from several authoritative sources.

6.4   Going off the Rails

In an article published in *The Sunday Times* on November 1, 1998, Gerald Corbett, at that time chief executive of Railtrack plc, invited readers to consider the economic interdependencies of the industry of which his enterprise formed a key part. In his piece, Mr. Corbett argued that investment and growth, "the privatized railways' great successes," risked being overshadowed by late trains and public relations disasters, and that both the successes and failures stemmed from the industry's "economic architecture." Privatization of the UK rail system had three key objectives:

- To cut the railways' government subsidy.
- To boost traffic: In 1995 there was no passenger or freight growth despite road congestion.
- To improve punctuality: A better service would encourage road users to switch to rail.

Somewhere in this set lie potentially the seeds of disaster. The objectives combine money (increasing revenue from greater traffic in order to compensate for reducing subsidy), quality (increasing service access while minimizing delays), and time (attracting revenue from enhanced service reliability). These are interdependent, just as the community of rail transport providers is interdependent. My goal was to analyze Mr. Corbett's ideas by employing a systemigram to capture his notion of "economic architecture," which putatively had been designed to address these objectives. The real issue was not whether this had proved to be the case (Corbett argued that it had patently failed), but whether anybody (or anything) could neutrally, objectively, and independently communicate on behalf of the community as a whole to its constituent enterprises, its members.

In the ensuing remarks, much of what that article contained is rehearsed in order to set as faithful a context as possible for exploring what is essentially an extended enterprise. Unless and until we are clear what this organizational landscape looks like, we cannot expect to perform any kind of thorough analysis, whether this be economically focused or otherwise, let alone reach conclusions as to the substance of improvements.

The pieces of this landscape are made up of four kinds of organization. First there are the twenty-five companies that operate the services for passengers and freight, known as the train-operating companies (TOCs). Each of these receives a steadily declining government subsidy and fare income that grows as traffic builds up. A second kind of organization is known as a rolling-stock company (ROSCo). The ROSCos own and overhaul the trains. A third element in the picture is Railtrack (now defunct and replaced by Network Rail, a government action precipitated by rail accidents with intolerably high fatalities), who owns and maintains the track, signals, and railway stations. Railtrack is in effect a supplier, of network capacity, to the TOCs. Finally, there is the government-appointed regulator of the system known as Opraf, the Office of Passenger Rail Franchise. Among its duties are the granting of licenses to the commercial elements in the rail system and the application of rewards/penalties for performance of these commercial service providers. The economic architecture to which the article refers is accounted for along the following lines.

The TOCs pay largely fixed charges to use Railtrack's lines and make fixed lease payments to the ROSCos. The fixed charge to use the lines was set to enable Railtrack to maintain and renew its network, upgrade all twenty-five hundred stations, and eliminate the big investment backlog. At the time of privatization, Railtrack's lines and signals were responsible for 65% of delays. Yet its income was largely fixed, giving it no incentive to improve punctuality. In response, the government gave it a performance regime with strong incentives to improve its network. The view was that the TOCs would not require incentives because if they performed badly, their fare income would fall as rail users retreated to their cars. This expresses confidence in self-regulation mechanisms.

Opraf introduced performance regimes but, according to Mr. Corbett's article, the incentives were weaker than for Railtrack. In 1997, Opraf paid £13 million net to the TOCs for performance, an average of £500,000 per TOC, but all of this was then paid by the TOCs to Railtrack under its performance regime. The economics of a typical TOC are something like what follows: Ticket sales produce 60% of revenue and the subsidy 40%. Of the outflow 40% goes to Railtrack, 18% to the ROSCos, 20% to the staff, and 18% to cover other costs. This leaves a 4% operating margin. In this situation, which levers will managers of such a TOC pull to improve profits?

First, the article argues, they will

> do everything they can to run more trains and attract more passengers. Almost all the extra income will pass through to the bottom line because most of the Railtrack, leasing and labor costs are fixed. Second, they will cut their costs because the state subsidy is falling by about 15% a year. This means sales must grow and costs must fall for the TOCs to stand still. Meanwhile, the theory was

that managers would still aim to run the trains on time because if they did not they would lose passengers.

Things have not turned out that way however.

The industry has grown faster than the architects of privatization imagined three years ago. Passenger miles have risen 16%, revenues by more and there are about 10% more trains on the network. The TOCs have cut costs—a 5% cut in jobs is typical. Their share prices have risen as higher sales and lower costs have come through to the bottom line. Railtrack has also performed, with delays caused by it and its contractors more than 40% down over three years. Infrastructure is now responsible for only about 45% of delays. There is more to be done, particularly on the Great Western lines, but progress has been made. Railtrack has more than doubled investment—to £1.25 billion last year and £1.45 billion this year. By the new year (1999), the British Rail backlog will have been almost eliminated, and the station upgrade program will be half finished. By 2001 Railtrack will have spent £1 billion more than the regulator assumed when setting its access charges in 1994. The state subsidy fell £285m this year to £1.6 billion and will be £926m in 2003–04. Privatization has delivered what its architects intended. But with success has come a problem—poor punctuality.

Railtrack, responding to its incentive regime, has cut delays caused by tracks and signals but the TOCs, with some exceptions, have not made similar progress. This is because their economic regime is potentially lethal for punctuality. Trains are added, jobs are cut, punctuality pressures mount, but the growth hides any loss in fares due to the poor performance. But the problem will not go away. As the subsidies fall, the pressure will build and a recession would intensify it. Some TOCs will be unable to keep investors and customers happy. The economic architecture will continue to drive them to actions that cause delays. What can be done? The industry could wait until 2003 when the TOCs' new franchise agreements start or it could seize the opportunity created by John Prescott.[9] New regulators and a new body, the Strategic Rail Authority, present the industry with a chance to correct anomalies in its architecture so that it can continue to grow and perform better. A new regime should include:

- Bigger incentives to make punctuality a real profit lever for the TOCs;
- An access charge related to sales to give Railtrack an incentive to encourage growth;

- Longer franchises to encourage TOCs to order new rolling stock;
- Government assurances that the industry will come through the regulatory review strong enough to fund investment;
- Existing subsidies to be realigned to provide incentives for desired outputs.

The privatized industry has much of which to be proud: good growth in passenger and freight traffic, billions of pounds of private capital in the sector, entrepreneurial managers with new ideas, steadily improving safety, a big rise in infrastructure and rolling-stock investment and a rescue for the Channel tunnel fast link. But the question remains: must travelers wait until 2003 for punctuality to be addressed? They deserve better.

How can we know whether Mr. Corbett is correct in his conclusions? After all, while one might appreciate his overview of the industry as a whole, pointing in some way or other to its systemic nature and the reality of its extended enterprise, he does and must have vested interests in the enterprise he leads, Railtrack plc. But this line would be true of any of the leaders of the other enterprises that make up Rail UK. In this case, how can we ever make sense of the totality, and which of several independent, possibly conflicting and certainly interdependent perspectives can we trust? This is not a trivial problem. And no matter how seductive a simple answer might look, for example, renationalization of Rail UK, we would be unwise to satisfy ourselves with grasping simplicity in the face of such complexity. Our response to this question is part of the systems thinking debate and at this point in time focuses on the validity of simultaneously tenable viewpoints and leveraging the validity of each, regardless of position, for all, in order to create a rich picture that achieves two important objectives: first, the encapsulation of these viewpoints (constituting the richness of the picture), and second, the neutralization of adversarial standpoints (through universal recognition of an essentially shared context).

Using the prose of Mr. Corbett's article as a source, and by concentrating on identifying the key elements and their interrelationships of the UK rail industry's extended enterprise, to which Corbett's words, thoughts, and concerns point, we created a systemigram (Figure 6.2). The systemigram affords a line of analysis that can easily be augmented by relevant expertise contained within the various elements and by case-hardened experience of the relationships between these elements. What is more, those served by this extended enterprise, the passenger and the freight owner, are kept in view. Indeed, these "stakeholders" have visibility into the complexities of the extended enterprise, the servant in the service they seek to obtain.

What principles can we infer from this exercise as an example of the value of the systemigram approach to architecting? Let us suggest six. First, the

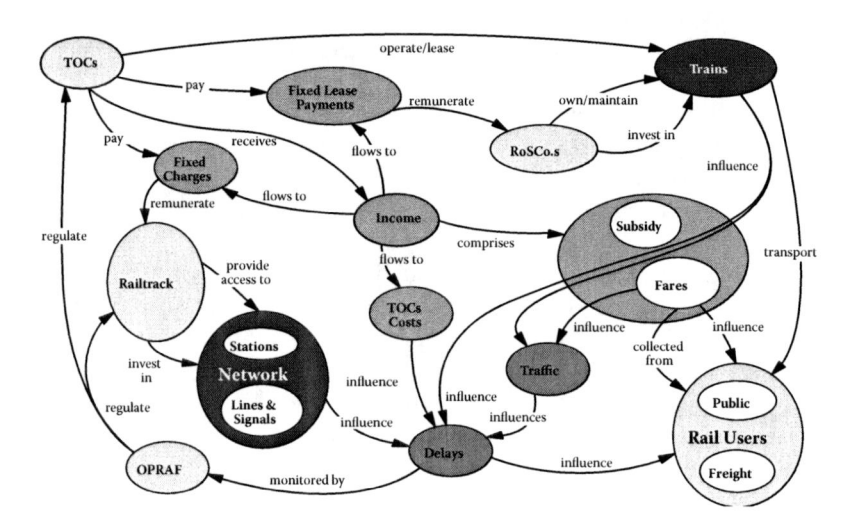

Figure 6.2 UK rail network.

notion that companies do work according to a well-laid-out system of processes is flawed. It is a system of companies that get work done, and the processes each uses necessarily interact with one another, across these notional corporate boundaries. Once a company accepts that much of the work it gets done on its interior critically depends on known (and unknown), modelable and unmodelable, modeled and unmodeled, it is breaking free of this flawed thinking and breaking through to new ways of working. The architecture we require is of the extended enterprise, and this necessarily involves multiple companies, multiple levels in the institutional hierarchy, multiple perspectives, and multiple agendas. The systemigram is tailor made for dealing with this phenomenon of multiple multiples.

Second, the extended enterprise architecture is a device for focusing on the need for and possibly means of integrating constituent services into a meaningful and purposeful whole. Without this integration driver the constituents will be forever operating as second best and their individual contributions counterproductive to the good of the whole, and inevitably their own good. The inherent integration features of a systemigram are fully supportive here.

Third, there are a variety of ways of capturing an enterprise architecture, but flow of money is as good as any, and one that gets people's attention primarily. Understanding the economic architecture of an (extended) enterprise is a vital prerequisite to effective change in the governance, security, and resilience of that enterprise. Systemigrams are network representations, and with the storyboarding technique a single systemigram can be portrayed as multiple networks. They therefore offer a natural medium for considering money flow.

Fourth, the multiplicity of viewpoints, from multiple agencies, is valuable because of the richness in perspective; it is also a potential hazard since it can so easily give rise to conflict. The goal is to leverage the diversity into a parsimony of focus—harmony. The systemigram simultaneously reveals the diversity in perspectives while maintaining a single objective "reality" that stakeholders can debate and ultimately attest and adhere to. It is a meeting point where consensus can be reached from rational evaluation of the multiplicity of viewpoints.

Fifth, the single objective reality might be anything but real, but it can be ideal, and the need is for stakeholders, constituents of the enterprise community, to collaboratively reengineer the ideal into an agreed upon reality, one that meets their diverse viewpoints, that can be strongly encouraged to cohere as focus on a single objective artifact is maintained.

Finally, all corporations are subject to buffeting—from fierce competitors, negligent suppliers, fickle investors, merciless regulators, and the invisible hand of the marketplace, whose movements are discernible seemingly only after the fact. Given this highly dynamic and uncertain state of affairs, it is inevitable that enterprises should seek to be resilient, that is, to be capable of resuming a normal operating mode following upheavals in the industrial landscape. Resilience should also mean being able to plan and prepare for such responses, lending the term *proactive agility*. But one corporation's resilience can mean another's disturbance, and if these are neighbors in the same extended enterprise, we have to move the notion of resilience one level up the scale and think in terms of the resilient extended enterprise. The challenge for corporations is to collaboratively design and build this resilience into the extended enterprises of which they are a part and, in doing so, piggyback off this collective effort and build for themselves a corporate-level resilience that satisfies both individual and collective good. The role that systemigrams can play in this multioptimization exercise is considerable.

We would argue that the systemigram approach can prevent corporations from bumping into one another, safeguarding against unwanted crashes, and keeping the collective corporate good very much on track.

6.5   *Normal Service Will Never Be Resumed*

We realize that making a television program is a complex affair requiring highly talented teams of people, perhaps some of whom will have egos, to work together under tight timescales and severe constraints in order to provide entertainment to a viewing public with no guarantee whatsoever that the audience will appreciate the result, let alone the effort that has gone into making the program. People would not do it if they did not love it. But that is only one side of the coin. The other side is getting the program into the homes of the viewers, without whom there is little point in the production exercise itself. Now this task, while perhaps not so glamorous or demanding, on the face of it at least, of huge creative talent, is not entirely a trivial one.

And in recent years it has been made vastly more complex with the advent of digital technology.

Once upon a time, and not that long ago, the viewer sat in his living room gazing at a simple TV screen, perched in the corner, tuned to Channel 1 in readiness for the sober announcement that *Play for Today* was about to commence. Today, using a single remote control unit (RCU), the viewer can surf hundreds of channels delivered by satellite and cable as well as 'off air', and talk back, through a phone line, to the makers of the program he/she watches, a program that competes on the same screen with information concerning other programs all displayed in neatly arrayed window frames. Now somebody has to take care of that line of business and manage those teams who make that business operate. All of a sudden the job of TV producer does not seem that big a deal, on the scale of things.

How did all of this change come about? Here is one perspective. In 1995, the UK government issued a White Paper and with it an unprecedented challenge to the broadcasting industry.[9] To quote from that paper:

> Digital broadcasting has enormous potential. If broadcasters decide to start using digital technology, they will be able to provide: many more channels, nationally and locally; a much better television picture or radio signal; new services (such as wide screen television and advanced teletext services); and much better reception, particularly for car radios and portable televisions. It will provide many people with their first experience of the full potential of the information superhighway. Using a telephone return link, it will allow home shopping and other interactive news, education and information services. And viewers will be able to browse through the channels available to plan their evening's entertainment.

The United Kingdom's leading broadcasting service at the time, none other than the BBC, responded with a suitably visionary statement from which we are able to extract:

> Over the next 20 years, digital technology will revolutionize the audio-visual services available to the home, school and workplace. It will transform the way programmes are made, and the way in which they are distributed through each of the main delivery systems—cable, satellite, telecoms and terrestrial. Above all it will deliver to audiences a range of new services—radio with CD quality sound; multi-channel television; pay-per-view; near-video-on-demand; interactive services such as home learning, home shopping and home banking; and, ultimately, true video-on-demand.[10]

Some promised land for us all, huh? But there is never a Moses around when you want one, and as I recall, though he led, he never actually made it to the land flowing with milk and honey.[11] I got a call that year from an old friend whose acquaintance I had met when I held the GEC Marconi chair of systems engineering at the University of Portsmouth, and from whose budget, in part, my costs of employment were being met. Ian Jenkins was at the time director of technology at the BBC, and he it was, if indeed it was anybody, who was being cast as Moses. Now, at the risk of prolonging a tedious biblical analogy, let me remind you that Moses did have some useful allies, and Dr. Jenkins knows better than anyone I know the importance of getting good allies, especially when the going gets tough.

The key question was: How can a corporation as historic, expert, and unrivaled in its business entirely reinvent one of its key assets, its delivery system, while at the same time figuring out the need to reinvent its content in line with all the upstarts whose competition was being acutely felt, especially in the areas of new programming, new services, and securing rights to premier sports events and blockbuster movies? The short answer is that it is not easy, but the problem is not going to go away. In its favor, the people running the BBC were changing because that is the nature and the increasing clock speed of the industry it is in, and if anybody can do it, the BBC can.

My involvement, as a consultant, was to turn over some hard ground, unearth some useful ideas, and present these as compellingly as one can to incredibly bright technical people working in a hugely talented artistic firmament. I took a systems approach, consulted widely, thought a lot, shared my thinking regularly, took notice of what people said, and determined to provide a disciplined framework for future collaborative activity, involving the BBC and a wide spread of companies with an eye out to digital broadcasting. In fact, it was my first taste of a genuine need for an extended enterprise to do business on behalf of real consumers as opposed to customers that were industries themselves, for example, consumers of large-scale software or capital goods such as telecommunication satellites or gas turbine engines.

As a result, I prepared a business process architecture (BPA) consisting of five core processes, each represented by a systemigram. The highest-level process, or business model as I called it, oversaw the others and drew together the BBC heritage with new regulation, the government's desire to sell off the analog spectrum, the need for investment (and the business case for this), the competition, and, above all, the viewing public. It was a neat picture (see Figure 6.3). It whetted the appetite and the audience cried for more. One of the really cool things at the time to consider was what was being referred to as interactive services, and so this theme was given prominence in one of the processes (see Figure 6.4).

But the exercise became more than simply designing a BPA and representing this as a family of systemigrams. It provided an opportunity to do something new, something different, and something that forms part of a crucial effort for systems engineering, and that is requirements management. My

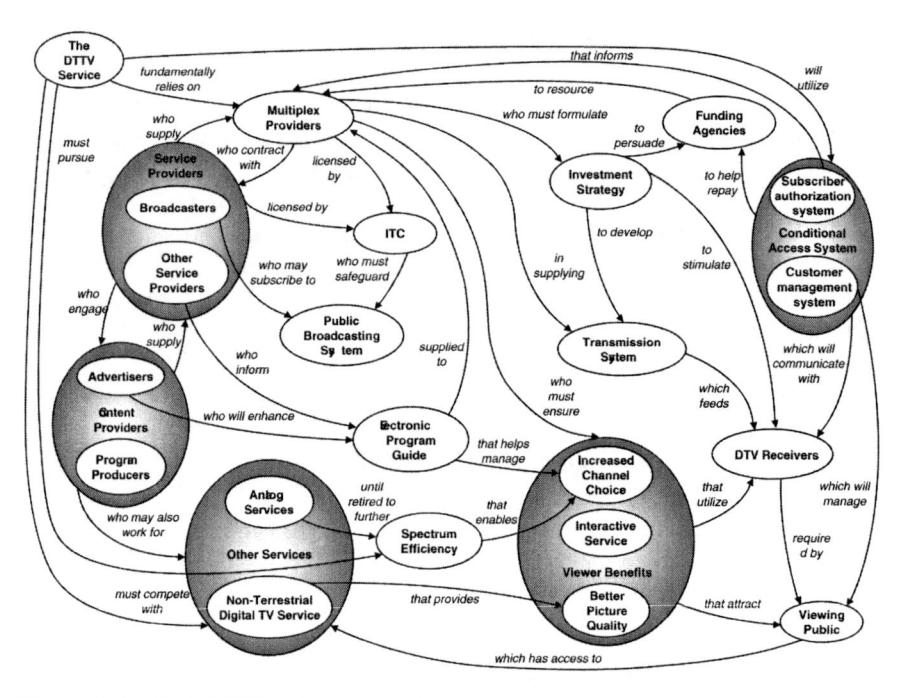

110   *Systems Thinking: Coping with 21st Century Problems*

Figure 6.3   Digital TV business model.

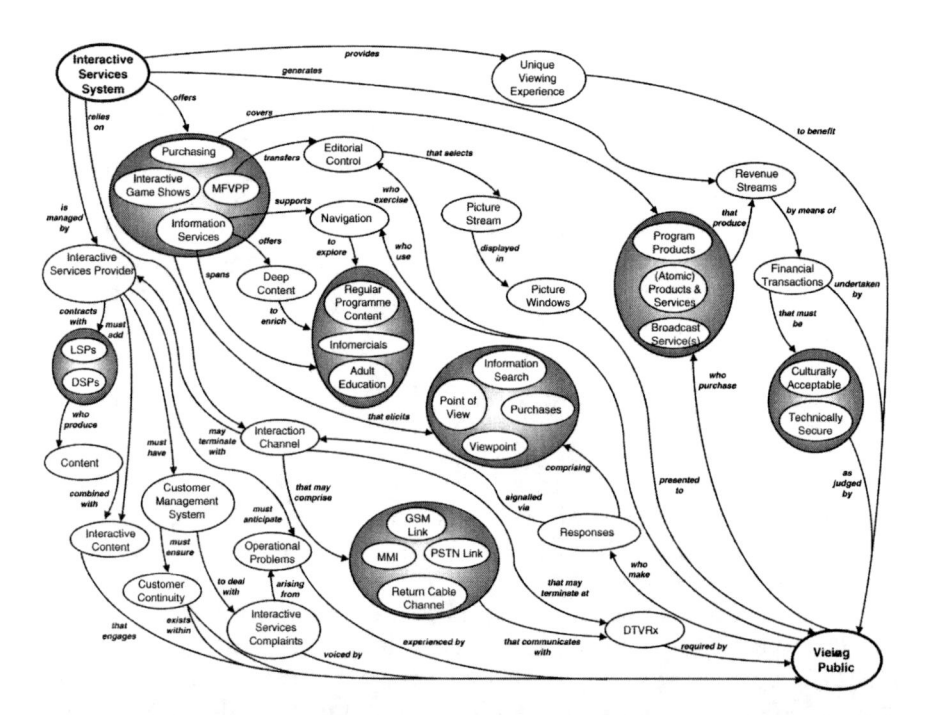

Figure 6.4   Digital TV interactive services.

involvement on behalf of the BBC in the preparation for the switch to digital TV was always seen as essential preparation, a necessary prior to the formal business of defining and articulating requirements from which would flow functional architectures and later physical architectures and thence the design, development, and procurement of hardware/software systems. My involvement was about as close to the front end as you could get, once the White Papers and their responses had been penned. My interest, as ever, was to know how this involvement would interface with the priors and the posteriors. What was the flow? What flowed between these efforts? What were the integrating artifacts?

Of course I did not have any answers to these questions, but that does not mean they are bad questions.[12] I resolved to invent the artifacts that would link this BPA systemigram family to the formal requirements management phase that was bound to follow this preparatory effort. I devised a matrix whose rows were the systemigrams, five in all, and whose columns were, as I saw them, the key concepts that could be found in two or more of the systemigrams. In the matrix cells I listed in bullet form what I saw as the major requirements being placed on these concepts (by their neighbors) or being placed by these concepts on their neighbors. The articulation of these requirements was drawn fairly naturally from the names of the links in the systemigrams. The result is surprisingly informative and a novel way of beginning the requirements definition process. As far as I was concerned, the systemigram approach proved amenable to a drill-down technique that was leading fairly naturally, with traceability all the way back to the White Papers, to the kinds of "shall" statements that systems engineers love. Considering how largely unobservable, at least for the vast majority of mankind, the initial set of shall statements has proved,[13] this recourse to traceability of strategic intent was not without merit.

6.6 Postlude

In summary, systemigrams have proved to be a novel medium for capturing strategic intent, in a way that prepares the ground for consensus building among diverse communicants and restores the inherent value of integrity, by which we mean the holding together of a system, the very basis for its stability, security, resilience, and agility. More than this they have provided a basis for system architecting, in terms of both enterprise integration (reliant on business process architecting) and technology systems development (reliant on requirements management).

As far as enterprise integration is concerned, the rules for employing systemigrams for architectural purposes are still in their formative stage. Interesting work has been done adapting Checkland's Soft Systems Methodology (SSM) to suit a business process architecting style.[14] The essence of this is threefold. First, human activity systems can be described in various ways, but taking a business (and technical) process approach has considerable virtue:

- It relates powerfully and compellingly to what people do.
- It yields a business process architecture (BPA) of the enterprise.
- It can accommodate an extended enterprise perspective, integrating the component BPAs into a system of systems view.
- It can provide a unique baseline from which to launch development projects.
- It animates an otherwise sterile library of processes into an active and adaptive portfolio of competence.
- It provides a benchmark for demonstrating and maturing competence of the enterprise.
- It affords a unique profile by which human skills, knowledge, and aptitudes can be successfully aligned with tasks.

Second, the BPA can be treated in exactly the same manner as a product architecture, relevant to life cycles, reviews, trade-offs, and simulation. In fact, the life cycle approach coupled with a top-down decomposition stepwise refinement of the processes yields views of the BPA that can be portrayed in systemigram form. Finally, the systemigram medium can provide a neutral and common environment for comparing and aligning the product, process, and enterprise architectures.

Concerning the communication of strategic intent and its linkage with technology systems development, there is scope for using families of systemigrams and the "children" that exist within any given systemigram—the scenes. A completed systemigram is not the end of the story. In fact, it is the basis for telling a story. The composer of the systemigram is now in a strong position, in spite of any illiteracy of the field being defined that he or she may have, simply because it takes considerable comprehension—of the original text and of building systems—to complete the systemigram. The story can be told in a variety of ways but all have the same generic format—to create a storyboard using carefully selected scenes that are subnets of the systemigram. This storyboarding helps to convey the message of the systemigram, together with the message that the author of the original text intended, to a wider audience.

Each scene represents a key part of the message, but by the same token it begins to tell a more detailed message that can only be amplified by having the right people listen to the systemigram story. So if there are, say, eight scenes, then in principle eight detailed messages can be generated, all at a lower level, but higher amount of detail than the systemigram. This drilling down can be continued for as long as required or until the messages begin to fail the original top-level requirements for original text for systemigram interpretation.

Chapter six: Systemigrams 113

In principle, this drill-down technique can be part of a requirements management approach in which the key concepts (nodes) are now increasingly circumscribed by relationships with peers, parents, and offspring. This approach was used successfully in the case study for the BBC in the arena of digital television.

Where do systemigrams go from here? To begin to answer that question we need to look at how they have fared in the United States with leading corporations and federal agencies, and, of equal importance, with the upcoming generation of industry and government leaders usefully sampled as master's students exposed to systems thinking.[15] If they had a successful, if somewhat painful, birth in twentieth-century Europe, how well might they mature across the pond in twenty-first-century America? A land facing the same stern tests that lay behind the thirty-fifth president's challenge.

6.7   Time to Think

1. Using the source text below, create a systemigram that follows the construction rules provided and conforms to the top-level requirements identified for the realization of a systemigram:

> The USAF Combat Strategy is to develop Constellation Net in support of Air Force CONOPS and Joint Vision 2020 that will lead to Joint Force Victory. The strategy must address information paralysis caused by stove piped networks that cause tribal warfare resulting in data-overloaded warriors, who operate in a camouflaged information environment, and lengthy kill-chains, which debilitates combat capability needed to secure Joint Force Victory. Strategy seeks to achieve predictable battle space awareness (PBA) and effects-based operations (EBO), which are required by USAF CONOPS and Joint Vision 2020, by developing horizontal integration to overcome stove piped networks. Strategy will adopt a staged process by means of:
>
> - defining network architectures for stovepiped networks and by developing technology and standards for information sharing across stovepiped networks and hence defining a network-of-network architecture;
> - achieving current system compliance with the network-of-network architecture; and,
> - acquiring next generation systems that will leverage off the network-of-network architecture by possessing in-built battle management peer-to-peer capability which achieves horizontal integration.

2. Using relevant sources, compose a two-thousand-word statement that captures the core issues surrounding the problem of illegal immigration into the United States. Construct a systemigram of your work and develop a storyboard of this with the specific intention of revealing these multiple perspectives and engaging a debate that will lead to consensus as to resolving the problem.
3. Using relevant sources, compose a two-thousand-word statement that captures the core issues surrounding the problems accompanying the disaster that befell New Orleans and nearby areas from Hurricane Katrina. Construct a systemigram of your work and develop a storyboard of this with the specific intention of revealing these multiple perspectives and engaging a debate that will lead to consensus as to the measures to be taken, at multiple levels in the nation's hierarchy, to better prepare for subsequent disasters similar in kind.
4. The systemigram in Figure 6.5 is a representation of Soft Systems Methodology. Analyze this in order to better understand the phases of this methodology and to create review points governed by suitable entry and exit criteria. Equipped with this deeper understanding of the methodology, suggest improvements to it that will deal with

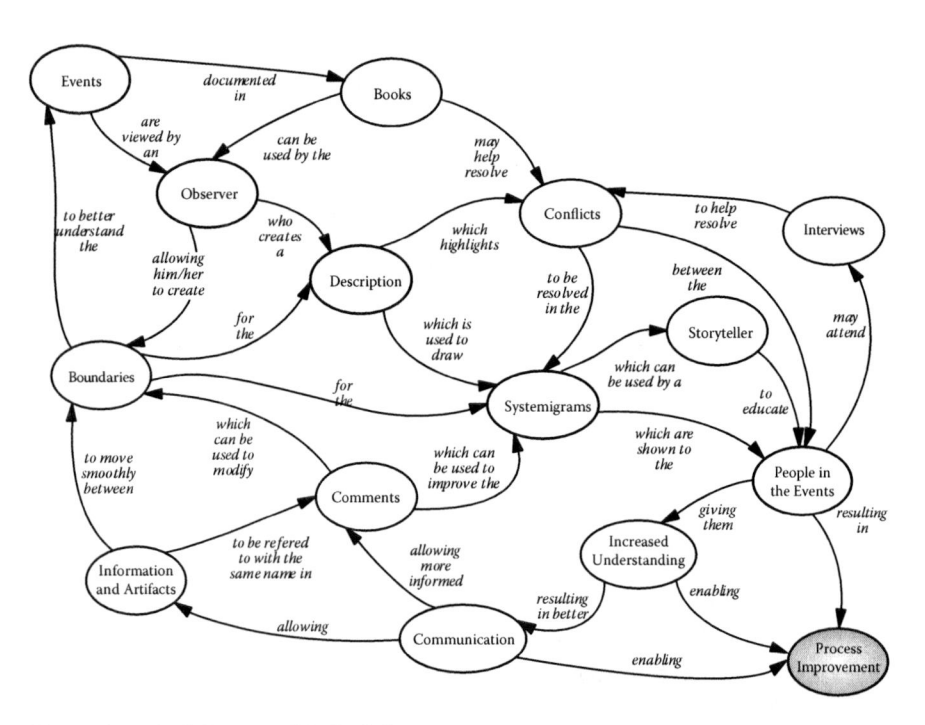

Figure 6.5 Soft Systems Methodology exercise.

misinformation, malicious compliance, antithetical viewpoints, and biases for executive action.

For your free copy of SystemiTool, go to www.boardmansauser.com for download instructions.

Endnotes

1. See, for example, http://www.sei.cmu.edu/pub/documents/93.reports/pdf/tr01.93.pdf.
2. Checkland, P., *Systems Thinking, Systems Practice*, Wiley, Hoboken, NJ, 1999, 175, Figure 9.
3. Novak, J. D., and A. J. Cañas, *The Theory Underlying Concept Maps and How to Construct Them*, Technical Report IHMC CmapTools 2006-01, Florida Institute for Human and Machine Cognition, Pensacola, www.ihmc.us.
4. The Ishikawa diagram (also fishbone diagram or cause-and-effect diagram) was introduced by Kaoru Ishikawa in 1982. See Ishikawa, K., *Introduction to Quality Control*, Springer, New York, 1991.
5. Senge, P., *The Fifth Discipline*, Currency Doubleday, New York, 1990.
6. An influence diagram (ID) (also called a decision network) is a compact graphical and mathematical representation of a decision situation. See Shachter, R. D., "Probabilistic Inference and Influence Diagrams," *Op. Res.*, 36, 589–604, 1988.
7. Rechtin, E., *Systems Architecting*, Prentice Hall, New York, 1990.
8. At the time Mr. Prescott was deputy prime minister and had responsibility for transport.
9. "Digital Terrestrial Broadcasting: The Governments Proposals," Cm 2946, HMSO, London, August 1995.
10. "Britain's Digital Opportunity," BBC publication, October 1995.
11. Deuteronomy 34.
12. "It will be a sad day for man when nobody is allowed to ask questions that do not have any answers." In Boulding, K., "General Systems Theory: The Skeleton of Science," *Manag. Sci.*, 2/3, 197–218, 1956.
13. Exodus 20.
14. Cole, A. J., Ph.D. thesis. "An Agent-Centered Method for Systemic Improvement of Business Processes." University of Portsmouth, Hampshire, United Kingdom, January 1998.
15. For example, the SDOE program from Stevens Institute of Technology.

chapter seven

Togetherness

7.1 Common Causes

The problems that appear in the world of work, often manifested in the technology we use to furnish this world, are essentially social and caused by a lack of togetherness or of bringing togetherness about. When the problems are genuinely social, that is, a property of the groupings of people in the world of work, it might be said that this lack of togetherness is caused by the inherent flawedness, some might say fallenness, of every individual, and that the systemic failure can be isolated to a particular individual (or part). Not real systemic failure in other words. Because parts in systems, like the fan blade in a gas turbine engine, are so closely coupled, tied together, a part failure leads to an engine failure, and the engine failure is systemic because the component failure cannot be isolated because that part has become much less of an individual piece and more an inseparable part of the greater whole.

But people systems should be as whole as our technological systems are designed and expected to be. Yet they are not. It is as though the fundamental individuality can never be subjugated to the good of the whole. But is this so? What of the firefighters of the FDNY on September 11?[1] These were individual heroes and true parts of a greater system, the human race. They were willing to die, and did so for the good of others in the system. There are many other such examples of sacrificial giving and loving, true partness in the cause of wholeness. But must it always take an emergency of historic proportions to bring this out? Can it not be played out in the ordinary stuff of the world of work?

In this chapter we are going to look at two distinct yet interrelated examples from this world of work, a world in which large numbers of individual workers seek to engage in collaboration in order to pursue purposeful activity united by a common cause. We propose to apply the systems thinking we have learned thus far, and especially the systemigram technique that we introduced in the previous chapter. Our goals are multiple. We genuinely want to apply systems thinking and see what success and failure this brings. We want to better understand the specific problems we encounter in these examples. In so doing we want also to explore what might be the underlying nature of these problems, which might somehow be more accessible to articulation and resolution by applied systems thinking.

We will also look at the proposed resolutions to the specific problems and examine just how well these might or might not work. Ours is a holistic approach. We do not use that word as an academic nicety or to be fanciful. Our aim is to be extremely practical and in that regard we are happy to be so judged.

7.2 An Intelligent Community?
7.2.1 The Way It Is

On April 12, 2005, Hanna Rosin, a *Washington Post* staff writer, gave both a credible and risible account of life in the community of U.S. intelligence agencies. Her opening anecdote, an office rumor, was used to capture both a symptom and a safeguard of the mayhem she observes in that community. It concerned the location of the office of John Negroponte, the first-ever director of national intelligence (DNI). The DNI faces many daunting challenges—courting foreign intelligence sources, for instance, and streamlining intelligence gathering to help prevent another massive terrorist attack—but in spy land the question that dominates discussions as Ms. Rosin has it is: Where will Negroponte's office be?

"If the president places him in CIA headquarters," says one former CIA official, "that will send the message that he's the boss now. If instead he's detailed to an alternative site in Tysons Corner, that would send the message either that he's irrelevant, or that the CIA's irrelevant, depending on whom you talk to. No one actually knows what the plan is, but the answer is beside the point. The real purpose of this office rumor is to keep alive the gossip and jockeying for power and endless squabbling that the new position was intended to end."

Sources and sharing are at the core of the DNI's brief. But whose sources and how to share are the bigger questions begged. As far as the National Security Agency (NSA) is concerned, its signal intelligence cannot be beat. But ask the CIA, and it will tell you its human intelligence is what counts. Which agency is more intelligent? And when you know the answer to that, help the poor DNI as he agonizes over "What do I do to change a culture conditioned by 'need to know' into one that lives for 'need to share'?"

In its final report, the September 11 commission called the system for sharing intelligence between agencies unacceptable, outmoded, and excessively secretive. The DNI is intended to get the agencies to stop hoarding and start sharing. But the early reports, according to Ms. Rosin's intelligence, do not look too hopeful. So far all the buzz has been about power struggles—DNI up, CIA down, Pentagon nervous—anything to give the fifteen agencies Negroponte oversees an excuse to give each other the silent treatment.

The intelligence world is a "community" only in the same sense as any high school. From the outside they are united by a common rival. But from the inside they are fractured into finely subdivided cliques that would not be caught in the same room together unless the DNI calls them into his office, wherever that is. Is the homeland secured by high school thinking and junior college pranks?

Spy land is populated by many tribes, and the majority view is these war among themselves with a fervor that should strictly be reserved for the common rival. A major breed is made up of the techno-geeks and a second by the 007s. Each side thinks the future of intelligence rests with it and the other side is for losers. Ms. Rosin quotes Robert Baer: "It's cubicle city. Computer guys, cryptographers. A bunch of people listening to inane telephone chatter for 45 minutes at a time. My God, it really puts you to sleep. Believe me, they don't have very exciting lives." That is his take on the National Security Agency, that big top-secret fortress at Fort Meade that is the headquarters of the techno-geeks. For their response, Ms. Rosin quotes James Bamford: "A CIA agent is someone who gets a lot of glory for intelligence collection, but 85 percent of intelligence comes from the NSA. Human intelligence never produced much useful information. And whatever they did produce was all compromised by Aldrich Ames and Bob Hanssen.[2] They never penetrated al Qaeda, and their intelligence on Iraq was marginal at best."

When these breeds are not at one another's throats, they are engaged in internal power struggles within their own ranks. Within the 007s the legendary spitting match between the CIA and the FBI continues to rage, ever more so now that the FBI is encroaching on foreign intelligence gathering. Here is some more of Ms. Rosin's invective:

> To the movie-going public they are both guys with trench coats who rough up the bad guys. But to each other they are different species, night and day, Jekyll and Hyde. A CIA case officer looks at the FBI agent and sees: a guy in Hush Puppies and a fake Burberry, clean-cut as a Mormon, never been to Paris or Morocco, never been far outside Fairfax. Every morning he gets in his Crown Vic and promptly clocks in. He's got some skills in hunting down bad guys, but he's also got a lawyer sitting on him all the time. Asking him to catch terrorists is like asking your kid's teacher to break up the local gangs.
>
> The FBI guy looks at the CIA guy and thinks: With a slight tick and shift in his history he'd be stealing cars in the Bronx. Gosh, he looks like he's been up a lot of nights in a row. Doesn't he own a razor? And how does he afford that place in Georgetown, not to mention those shoes?

It may be just another line from another movie but it is a telling one: When one character in *The Good Shepherd* says to the CIA agent Edward Wilson (played by Matt Damon) "Let me ask you something ... we Italians, we got our families, and we got the church; the Irish they have the homeland, Jews their tradition; even the niggers, they got their music. What about you people, Mr.

Wilson, what do you have?" Mr. Wilson replies, "The United States of America, and the rest of you are just visiting." Hardly ignorable. In fact, memorable.

Presidents back to Gerald Ford have tried to gather the various intelligence branches into one big happy family. One government Web site[3] hails the power of cooperation and shows seemingly happy colleagues working shoulder to shoulder. But those who know better sigh, like the principal facing the same old boys in his office. "It's not a problem that can be solved," says James Pike, director of www.globalsecurity.org. "It's just a process that has to be managed."

7.2.2   The Way It Must Be, OK?

As much as Ms. Rosin might be believed, she is still for many people first and foremost a newspaper reporter. People choose to believe what they want when they read newspapers. But it must surely be an inconspicuous minority that clings to disbelief in the Robb-Silberman report, the deliverable from the Commission on Intelligence Capabilities of the United States regarding weapons of mass destruction.[4] In the commission's transmittal letter to the president of the United States dated March 31, 2005, we find this:

> We conclude that the Intelligence Community was dead wrong in almost all of its pre-war judgments about Iraq's weapons of mass destruction. This was a major intelligence failure. Its principal causes were the Intelligence Community's inability to collect good information about Iraq's WMD programs, serious errors in analyzing what information it could gather, and a failure to make clear just how much of its analysis was based on assumptions, rather than good evidence.

Dead wrong! That is unignorable. The problem the commission has, though, it seems to us, is that their report of over six hundred pages is decidedly ignorable simply because it is too much information. Is there some way to make it unignorable? Is there a way that gets us from the "dead wrong" conclusion to the "live right" code as effortlessly and as accurately as possible?

A classic technique is to capture the essence of a report, as of course the transmittal letter does for that particular report. An executive summary gets us a little farther. Our interest is in capturing strategic intent and ensuring the governance of tactical operations that fulfills this intent. While not ignoring the rest of the report there is a part of it, Chapter 6, that perhaps deserves a sharper focus. It is entitled "Leadership and Management: Forging an Integrated Intelligence Community."

Of course forging requires heat and melting, and who cares to get burned and lose their identity in the process? It is true, some people do have to be dragged kicking and screaming to their blessings. But who can tell that blessings there shall be when people's experiences have largely been to the contrary?

Chapter seven: Togetherness

The problem is that it is not a trivial problem to define, added to which it is not an easy solution to produce, and if one could find a solution, can it be one whose implementation satisfactorily deals with the problem? At the risk of incurring paralysis by analysis we do well at the very least to acknowledge this "wickedness."

In our attempts to apply systems thinking to this problematique, we created summary text based largely on a validated comprehension of Chapter 6. It is as follows:

> The intelligence community (IC) comprises badly equipped and badly organized agencies, which includes assuming an analytical community that is too slow communicating with policy makers. The agencies use traditional techniques that have declining utility against increasingly elusive and diffuse threats to destroy the United States and its allies. The IC suffers from institutional incapacitation created by stovepiped intelligence expertise that results from fragmentation of badly equipped and badly organized agencies.
>
> The IC has a responsibility to safeguard the *United States* and its allies, and this can only be achieved by transformation of the IC, which will require a *four*-part, *one*-whole strategy: focus on missions, integration leadership, HR transformation, and information accessibility.
>
> The focus on missions will be supported by target development boards and will result in an integrated end-to-end collection enterprise. Integration leadership will require integration intelligence strategies and planning, programming budgeting and execution (PPBE), and should produce fused domestic and foreign intelligence enterprises and an end-to-end budgetary process. HR transformation will utilize a common personnel performance evaluation and compensation plan and should considerably enhance talent attractiveness, drawing in outsiders with relevant skills and rewarding them on merit, thereby improving the agencies' resources. Information accessibility must overcome the stovepipes and the gaps and uncertainties in intelligence and should produce a community that can rapidly reconfigure itself to respond to crises.
>
> This transformation of the IC will lead to integrated intelligence and so help safeguard the *United States* and its allies.

This text gets at some of the ailments of the IC and integrates treatment with those ailments in the context of what it is the IC is supposed to achieve on behalf of the United States and its allies. As a result, we produced the systemigram in Figure 7.1.

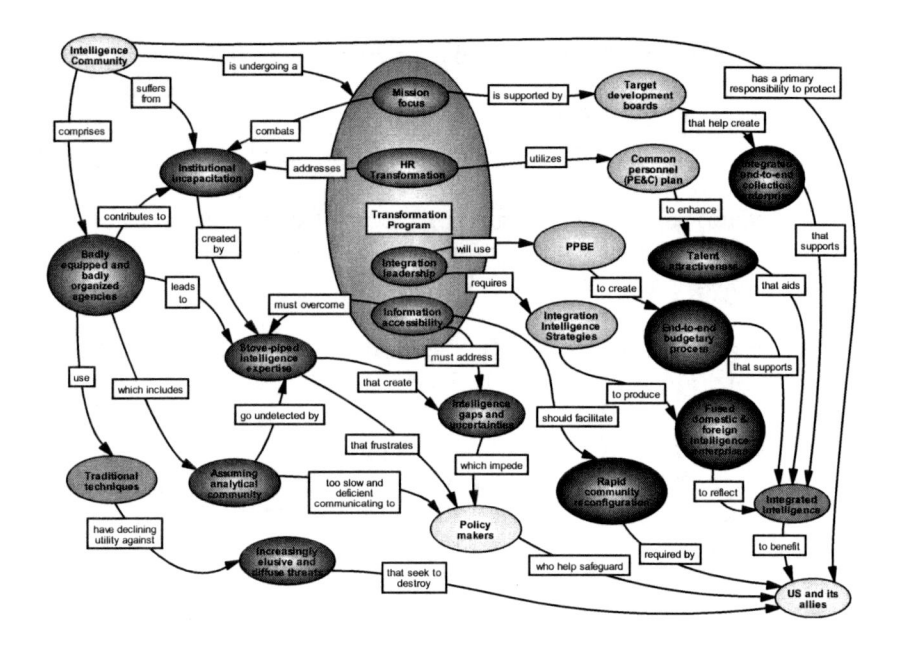

Figure 7.1 Intel community.

This work has been exhibited to leading experts—thinkers, analysts, strategists, and advisors in and to the IC. That exhibition gave us our validation, and the dialogue it supported gave us a systemic means of getting a handle on the wicked problem itself. Our approach is in line with the mantra "It's not a problem that can be solved, it's just a process that has to be managed." We believe we are providing management support to that process—of self-analysis, in a context of reasonable self-esteem and recognition to build on success for higher successes—those of the community and those it seeks to serve.

7.2.3 Making Sense of Togetherness: 1

Our motivation in this chapter is not just to show how systemigrams can be used but to address the very nature of the problem situations that cause us to confront their intractability, and paradoxically to insist that this suggests a means for managing the process if not solving the problem. Some might argue that the term *intelligence community* is as much an oxymoron as military intelligence. Rather than consigning this remark to the cheap-shot category, we want to say there is a paradox with the term and we will leverage this paradox to find wisdom from above, trusting in systems thinking as a means of finding this wisdom, and staying positive as regards a feasible approach.

In that spirit we inspect the above systemigram not as a solution in and of itself but as a lens into the problem and how it is proposed to address it, and therefore as a valid child of the parent report.

Our first observation is that the IC is "separated" from the United States and its allies. But this separation should not simply be interpreted as a negative, albeit many would say—as indeed the report does say—that is the way it really is. Our intention is to reflect the prime real estate value of these two concepts (verities in the real world) since they are the beginning and end of our design—the alpha and omega. The space that separates them is filled with ills, goals, and programs. It is good to reflect that this is what separates them—the ailments of the IC as it is, the goals to be reached, and the programs that will form the bridge. How well have we filled this space?

To answer this question, we make a second observation concerning stovepiped expertise. Is this inevitably a consequence of badly organized agencies? Or is it a consequence of the "need to know" culture? And are these two questions the same? That is, does the way that agencies are organized serve the purpose of need to know and the badness of this organization lead inexorably to stovepiped expertise? If this is true, then information accessibility alone cannot overcome stovepiped expertise—it is simply asking too much. While information accessibility must be regarded as a desirable effect, the question is begged, What will cause this effect?

At this point we might observe that the diagram is failing us. Much is made of the need for integration, and we could easily create scenes that show this emphasis, focusing on the integrated intelligence bubble and the four bubbles that input to this, such as fused domestic and foreign intelligence enterprises (there is another tough word to swallow—*fused*—if you are part of the fusion). But we do not see how this is to be achieved. The day is saved by the realization that it is institutional incapacitation that needs to be overcome, and this by the goal of integrated end-to-end collection enterprise served by the mechanism of target development boards. The diagram has not failed us—in this instance. However, it does fail us in another—that of assuming analytical community.

The diagram does not serve us as well as it could, or at least a different one could—another child from the same parent, as regards this key impediment. What we can say is that there is a direct link between the assuming analytical community and the policy makers. This affirms what the report declares—that the analyst is the voice of the IC. The job of the analyst is to tell policy makers what they know, what they do not know, what they think, and why. What further impedes the policy makers are intelligence gaps, so that between these and the assuming analytical community this is a double whammy for the policy guys. Of course, both these ailments are connected with stovepiped expertise that is by its nature unable to detect the existence of the assuming analytical community. The report recommends that they engage in competitive analysis, not by hiding information from one another but rather by suspending their belief in what they know, taking on trust that others know what they know. The report suggests that mission managers are key to this culture change, but the diagram does not say this as emphatically as it ought. However, this is not a total failure if inspection of the diagram reveals this missing emphasis.

Our final observation is something we have already touched upon: the report is the parent; systemigrams are the children. A family of systemigrams owes its lineage to the report. Siblings squabble and it is not a disaster that any single systemigram fails to convey all—the family as a whole might do better. All of this is proof, if proof were needed, that it is never easy to build systems unless the specification is crystal clear and the priorities well known. But we are moving into an era where specifications are nonexistent and priorities are ephemeral. Our response? To build systems that can build themselves. This takes us into the arena of self-organizing systems or, as some would have them, system of systems (SoS). This is of interest to the Department of Defense (DoD), to whose world we now turn.

7.3 The Madder of All Wars

September 11, 2001, is a landmark in human history. For those that argue that it changed nothing, let them at least agree that it signaled a new reality for those who had thought differently about the world. For the U.S. military the events of 9/11 confirmed what was already known and gave special impetus to their ongoing transformation to deal with an asymmetric threat and to support the drive for Joint Force Victory. In this section we wish to concentrate our attention on the USAF's future capabilities and find ways of supporting its goal to achieve these in the context of a changed world.

7.3.1 Future Capabilities

The Air Force's top officials told the Senate Appropriations Defense Subcommittee that America's Air Force was ready, willing, and able to defend the nation's skies. The following are excerpts from the joint statement in 2004 of Secretary James G. Roche and Chief of Staff Gen. John P. Jumper:[5]

> We will adopt service concepts and capabilities that support the joint construct and capitalize on our core competencies. To sustain our dominance, we develop professional Airmen, invest in warfighting technology, and integrate our people and systems together to produce decisive joint warfighting capabilities....
>
> The Air Force has written six CONOPS that support capabilities-based planning and the joint vision of combat operations. The CONOPS help analyze the span of joint tasks we may be asked to perform and define the effects we can produce. Most important, they help us identify the capabilities an expeditionary force will

need to accomplish its mission, creating a framework that enables us to shape our portfolio.

Homeland Security CONOPS leverages Air Force capabilities with joint and interagency efforts to prevent, protect, and respond to threats against our homeland—within or beyond U.S. territories.

Space and Command, Control, Communications, Computers, Intelligence, Surveillance, and Reconnaissance CONOPS (Space and C4ISR) harnesses the integration of manned, unmanned, and space systems to provide persistent situation awareness and executable decision-quality information to the JFC.

Global Mobility CONOPS provides Combatant Commanders with the planning, command and control, and operations capabilities to enable timely and effective projection, employment, and sustainment of U.S. power in support of U.S. global interests—precision delivery for operational effect.

Global Strike CONOPS employs joint power-projection capabilities to engage anti-access and high-value targets, gain access to denied battlespace, and maintain battlespace access for required joint/coalition follow-on operations.

Global Persistent Attack CONOPS provides a spectrum of capabilities from major combat to peacekeeping and sustainment operations. Global Persistent Attack assumes that once access conditions are established (i.e. through Global Strike), there will be a need for persistent and sustained operations to maintain air, space, and information dominance.

Nuclear Response CONOPS provides the deterrent "umbrella" under which conventional forces operate, and, if deterrence fails, avails a scalable response.

Through capabilities-based planning, the Air Force will continue to invest in our core competency of bringing technology to the warfighter that will maintain our technical advantage and update our air and space capabilities. We need to field capabilities that allow us to reduce the time required to find, fix, track and target fleeting and mobile targets and other hostile forces. One system that addresses this operational shortfall is the F/A-22 Raptor. In addition to its contributions to obtaining and sustaining air dominance, the F/A-22 will allow all weather, stealthy, precision strike 24 hours a day, and will counter existing and emerging threats, such as advanced surface-to-air missiles, cruise missiles, and time sensitive and emerging targets, including mobile targets, that our legacy systems cannot. The F/A-22 is in low-rate initial production and has begun Phase I of its operational testing. It is on track for initial operational capability in 2005. A complementary capability is provided by the F-35 Joint Strike Fighter, providing sustainable, focused close air support, and interservice and coalition commonality.

We also recognize that operational shortfalls exist early in the kill chain and are applying technologies to fill those gaps. A robust command, control, and sensor portfolio combining both space and airborne systems, along with seamless real-time communications, will provide additional critical capabilities that address this shortfall while supporting the Joint Operational Concept of full spectrum dominance. Program definition and risk reduction efforts are moving us towards C4ISR and Battle Management capabilities with shorter cycle times. The JFC will be able to respond to fleeting opportunities with near-real time information and will be able to bring to bear kill-chain assets against the enemy. Additionally, in this world of proliferating cruise missile technology, our work on improving our C4ISR capabilities—including airborne Active Electronically Scanned Array or AESA radar technology—could pay large dividends, playing a significant role in America's defense against these and other threats. To create this robust command and control network, we will need a flexible and digital multi-service communications capability. We are well on our way in defining the architecture to make it a reality. The capabilities we are pursuing directly support the Department's transformational system of interoperable joint C4ISR.

There is a need for a globally interconnected capability that collects, processes, stores, disseminates, and manages information on demand to warfighters, policy makers, and support people. The C2 Constellation, our capstone concept for achieving the integration of air and space operations, includes these concepts and the future capabilities of the Global Information Grid, Net Centric Enterprise Services, Transformational Communications, the Joint Tactical Radio System, and airborne Command, Control, and Communication assets, among others.

Our view of the above is articulated in the systemigram in Figure 7.2. We defer commentary on this until we have had a chance to look inside one of the key bubbles in this systemigram—C2 Constellation.

7.3.2 A New Constellation

The Air Force's contribution to the overarching concept for warfighting operations is the C2 Constellation—the Air Force's components to the Global Information Grid [GIG]. The C2 Constellation is a family of C4ISR systems sharing horizontally and vertically integrated information through machine-to-machine

Chapter seven: Togetherness 127

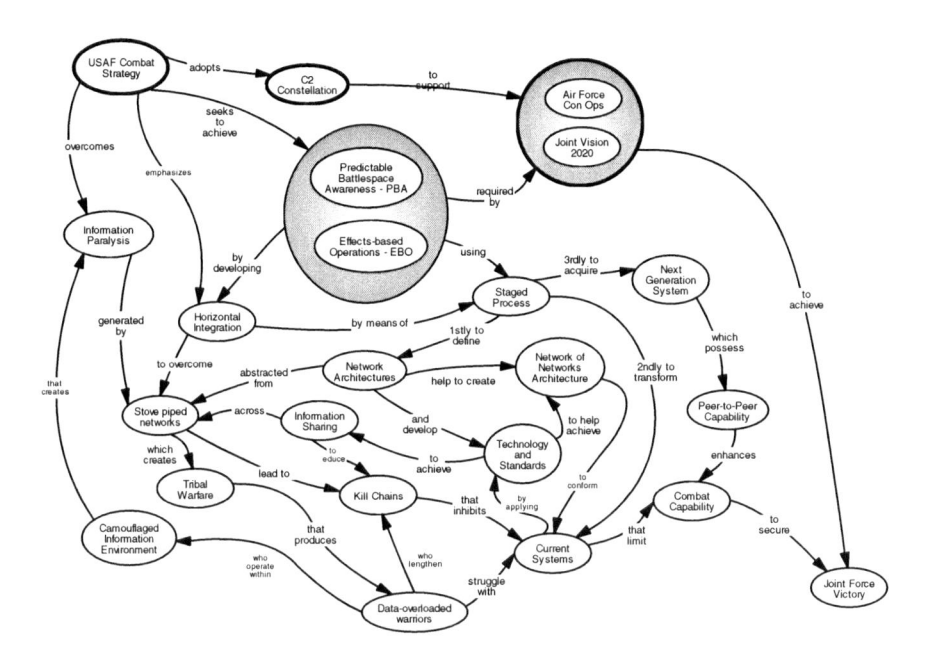

Figure 7.2   USAF combat strategy.

conversations enabled by a peer-based network of sensors, command centers and shooters. Both an operational construct and an architectural framework, it guides the Air Force's development of people, processes, and technology toward network-centric operations.

Key network-centric operation elements of the C2 Constellation include the various platforms and sensors the Air Force provides to the Joint Force Commander and key programs that support command centers such as the Air and Space Operations Center and the Distributed Common Ground Segment. Underpinning programs within the AOC, such as the Theater Battle Management Core System, already serve as the joint standard for air operations planning and execution, and we are continuing to migrate these systems to a more modern, web-enabled architecture.

The Air Force provides transport and computing layer components of the overall DoD GIG through Constellation Net, the communications network—air, space, and terrestrial—that facilitates free flow of information, rapidly accessible to our warfighters. The Air Force portion of GIG Bandwidth Expansion provides expanded terrestrial service at key Air Force bases globally. The Joint Tactical Radio System is essential to our vision for an improved airborne network, which expands genuine network operations to the airborne platforms. With the installation of Family of Advanced Beyond line of sight Terminals on additional aircraft, such as AWACS, JSTARS and Global Hawk, we will have the capability to extend our airborne network to all reaches of the globe. Finally, the Air Force is responsible for a large portion of the space segment communication evolution, including deployment of the Advanced EHF, Wideband Gapfiller System, and the Transformational Satellite program....

Early discussion of architecture and network-centric requirements are driving early direction and management decisions for key programs at the DoD level. Facing the need to recapitalize its aging DCGS, the Air Force is working to eliminate stovepiped intelligence processes and bridge information divides between the Joint operational and intelligence communities through the Block 10.2 Multi-INT (multi-intelligence) Core. As part of this effort, the Air Force approach develops an open-architecture-based DCGS Integrated Backbone for the broader DoD DCGS modernization effort, designed to be inherently joint and interoperable. Formerly referred to as the Multi-Sensor Command and Control Constellation, or MC2C, in 2003 the concept underwent a name change, becoming simply the C2 Constellation. The Air Staff opted for the new name in part to avoid confusion some perceived between MC2C and MC2A, the Multi-Sensor Command and Control Aircraft. The constellation itself represents the effort to fulfill a vision put forth by Air Force Chief of Staff Gen. John P. Jumper. He wants to see a fully connected array of land-, platform-, and space-based sensors that use common standards and communication protocols to relay information automatically in what he refers to as machine-to-machine interfaces [sic].

The constellation is really a nontraditional management and program execution approach to providing vastly improved command, control, computers, communications, intelligence, surveillance, and reconnaissance capabilities to operators. Quarterly integration council review meetings, the first of which was held in December, are prime examples of this nontraditional approach. In these sessions, program managers don't demonstrate the standard "quad charts," which track cost, schedule, and performance. Rather, they demonstrate the interaction between their nodes and others in the constellation. It isn't an aircraft, advanced radar or even a sophisticated software program—although all play key roles—that will ultimately define the success of what is now referred to as the Command and Control, or C2, Constellation. Common standards, new business rules that stress openness, and a robust modeling and simulation effort are the main ingredients, according to the deputy C2 Constellation system program

Chapter seven: Togetherness

director. The RFPs will contain common contract language. This with full knowledge that their bid must fit in a constellation construct. That's where the new business rules come into play. By reducing proprietary coding and requiring contractors to "make open" the work they've done for the government, so that it can be shared and built upon, ESC hopes to increase efficiency and achieve true integration. Initially, ESC expects some contractors to balk at these new rules. This is difficult for industry because their intellectual property has been very lucrative. Sharing that property with other contractors tends to fly in the face of their bottom-line return on investment. Still, ESC doesn't intend to turn a deaf ear to contractor concerns. The Center is looking at ways of providing contract incentives for those who willingly agree to work this way.

The C2 Constellation Program Office itself doesn't have as much money to spend as it originally expected. Anticipating about $100 million in Fiscal Year 2003, they received only about a fifth of that. Part of the reason for this was the inherent difficulty of explaining the need to fund something that is not, itself, a specific program. The constellation is very difficult to articulate. Nevertheless, the office has developed a plan. Part of the plan is to do much of the architecture development and systems engineering work, which would have been done on contract, in-house. The other part is to build on existing integration work already being done contractually for programs that make up the current nodes on the constellation. Part of the money that is available will also be spent on upgrading and continuing to test with the Paul Revere, the 707 testbed jointly operated by ESC and MIT's Lincoln Laboratory. The Paul Revere helps integrate air and ground battle management C2. Combined with the C2 Enterprise Integration Facility, which allows ESC to wring out a lot of connectivity and integration issues without the expense of flying, the Paul Revere lets us really test constellation concepts, especially in exercises such as JEFX.

Our comprehension of C2 Constellation is portrayed in the systemigram in Figure 7.3. Now that our brief sojourn of the USAF's new world has ended, and we have concluded this by drawing two important maps of this world, we are now in a position to make sense of the togetherness of which the USAF speaks—a togetherness of the United States and its allies, of the branches of the military, of people and systems, and of legacy and new systems—by any measure a nontrivial challenge.

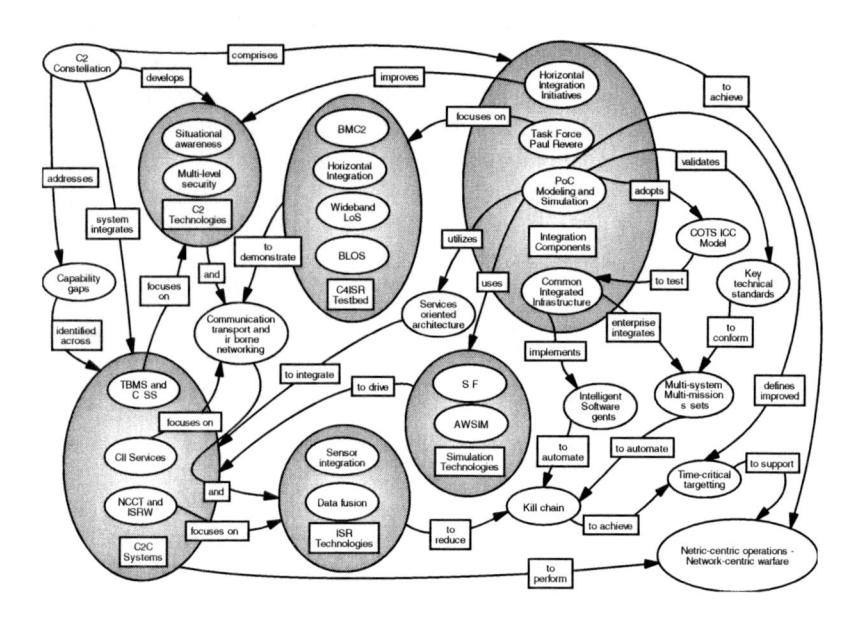

Figure 7.3 C2 Constellation.

7.3.3 Making Sense of Togetherness: 2

The USAF combat strategy systemigram has many similarities with the intelligence community systemigram. It has a similarity of concerns (stovepipes and information paralysis) and of the means to address these concerns (information sharing and horizontal integration). This similarity has good and bad points. A good point is that we can be convinced that this ailment is not uncommon and therefore there may be some more fundamental concern that can be addressed, and whatever resolution we can conceive at this more fundamental level can be suitably tailored to the specific circumstances in which it occurs. A bad point is that we might be tempted to see this problem everywhere, and in places where it is not, and get into a rut in insisting that systemigrams have this feature present whether it is applicable or not.

What, in these two cases, might be a reason for this similarity? The fact that fifteen of the twenty-seven intelligence agencies are part of the DoD? If so, then the togetherness challenges that are being set are the correct ones, and of epic proportion.

A second similarity between the two systemigrams concerns the fundamental approach adopted to these problems of stovepipes and information paralysis by both communities—that of transformation. Notwithstanding this similarity there does appear to be an important distinction in the approach to change. For the IC it is based on leadership, with systems being designed in support of this "soft" target. For the USAF it is the classic hardware approach—a phased transformation predicated on definition, conformation, and acquisition. This staged process firstly defines the architectures of the stovepiped networks, followed by the conformation of current systems in accord with the network of networks architectures generated from these definitions, and is concluded by acquisition of next-generation systems that will provide the much coveted peer-to-peer capability that epitomizes horizontal integration.

It should come as no surprise that the traditional paradigm of hardware-oriented thinking should be to the fore. But the distinction we find should allow for constructive criticism, for either case. The challenge we face today is bigger than simply systems, or for that matter systems and people. It is about how we think and how the way we have thought has let us down—badly. It is about having the courage to admit this and to hope that in the future we can think differently, not letting systems dominate our thinking but for a change thinking about systems (and people) in a way that overthrows this domination.

Our final comments on the USAF combat strategy systemigram are related to its spatial characteristics and the proximity or otherwise of key concepts. The C2 Constellation bubble is quite far removed from the kill chains bubble, and yet the whole raison d'etre for C2 Constellation is to shorten kill chains. Wouldn't it be tragic if this key goal somehow got overlooked in the process of building the solution to the problem? This would not be for the first time in the hardware-oriented thinking paradigm, especially when the technical challenge is immense taking into account huge egos, big budgets, and long timescales. These are all strong influences that accelerate the phenomenon of solutions creep. This separation is because of the C2 program. Logically you need a program to turn a goal into an achievement. But with a systemigram you not only get that logic but also get the warning that the program can get in the way of achieving the goal, sometimes unintentionally, hopefully not as a result of enemy action.

On the subject of enemy, the enemy of the lengthy kill chains is the stovepiped networks. How ironic that the very systems that the USAF currently has have become an enemy of what it is now trying to achieve. But this identity switch is perhaps a sign of the times: of the changing environment, of the asymmetric threat, of the penalty of legacy. But we need to ask, yet again, where do these stovepiped networks come from? Are they accidents of birth? Or is there some bad organization at work creating them? Maybe the process—definition, conformation, and acquisition—will work. But will this mean we do not learn the lesson of how this enemy came to be? And if we miss this lesson will it reemerge in the new set of systems (or system of systems)?

Some good news. The next-generation systems are in close proximity to the containment node with the CONOPS and Joint Vision 2020 bubbles. This is a good thing. These documents and their doctrines drive the next-generation systems. If this doctrine changes, so must these systems. If that is a problem for the system developers we had better know so, why this is, and what can be done about it. Finally, the horizontal integration bubble sits vertically above the stovepiped networks it is intended to overcome. This is sweet irony.

7.4 Principles for Togetherness

To close this chapter, we are offering a set of principles we call cohabits. They are so called because their practice must recognize the need for partness by individuals and the essential belongingness of those individuals to larger groupings, of people systems, to which they harmoniously render their work contribution. Likewise, these cohabits affirm the responsibilities of a group, as a whole, to safeguard the individuality of the parts and to encourage appropriate interdependency of those parts, which then safeguards the integrity of the whole, thereby increasing its attractiveness to the parts to pursue wholehearted belongingness.

So cohabits are the antithesis of the me-habits. They discourage and repulse egocentricity in favor of familial or genuine societal needs, which then leads naturally to what every individual desires, an enhancement of acceptable self-worth, self-esteem, and self-confidence. It leads there in ways more desirably and more rewardingly than the egocentric motivations of self-preservation or personal aggrandizement could ever achieve. Much more.

A cohabit therefore is a piece of advice or guidance to help you get your mind clear, your thinking straight, or your actions aligned with the idea that the world of work has lots of people in it whose very being and courses of action are strongly interdependent with your own existence and contributions. If you accept, adopt, and adapt this advice, it will become a habit that should enable you to get alongside and get along with others in your world. It should become a cohabit. We have tried to draw out from the words of this book and using experiences from our world of work an initial collection of cohabits. Below we have provided a top ten for each of these headings: coexistence, cooperation, and coeducation.

7.4.1 Coexistence

1. You are not alone. You might feel that you are and all the circumstances of life may strongly suggest to you being alone, but you are not. You may even have a job that requires you and you alone to carry it out, which is something that reinforces isolation. Nevertheless, somebody somewhere out there needs you. And not just the output of what you do. They need you. Affirm to yourself, regardless of "facts" and feelings, that you are a coexister.
2. Your very existence enhances the lives of others, and when you affirm yourself to be a coexister your own existence is enhanced as well. It begins to assume the nature of coexistence.

Chapter seven: Togetherness   133

3. You have the awesome responsibility in your work to enhance the work of others. When you recognize this opportunity and seize it, the value of what you do is not solely measured by what you do but also by what others do.
4. Now that you accept you are not alone, and more than this that your very existence is enhanced by the coexistence of others, you will be more readily able to accept that your viewpoint on the world of work is one of many. A coexister accepts that all of these viewpoints are simultaneously tenable even if that leaves you with a bunch of apparent paradoxes. Live with the paradoxes and go with the simultaneously tenable viewpoints (STVs).
5. There is a way of making sense of the STVs and banishing the apparent paradoxes. It is the cohabits way of course. The way is a search for an essentially shared context, the objective reality on which the STVs subtend; after all, a viewpoint, however real to the viewer, is ultimately subjective. The way is also one of defeating adversarial conflict, easily produced by the STVs, through a process of recognizing and exploiting richness in variety.
6. Just as you are not alone, neither is a firm. A concrete example of inclusivity for a firm is the alliance: a systemic union of two or more firms intended to achieve a common purpose facilitated by common principles of endeavor.
7. Alliances do not come free: they demand and consume management attention and may even turn into black holes. When forming and managing alliances, watch out for free riders.
8. Alliances are more than their mechanics, for example, legal structure, governance, gain sharing, and exit strategies. They require strategic logic for their creation and maintenance, a logic that must draw on cohabits; me-habits will saddle you with burdensome overheads instead of affording a vehicle for slick transport.
9. Do not be satisfied by the cosmetics of business when you need to look deeper into its being, its essence, what makes it unique. To make sense of business in the cohabiting world, you need to know its DNA, what allows it to transact with the other living corporate organisms that taken together create new forms of life and richer values in the market place. Process architectures give access to the DNA of business, the genetics of business competence.
10. A silo mentality or mindset in business fuels the war on costs. One department gets the cheapest price from a supplier, and another department picks up the costs of having to fix problems with what that supplier provides. These two departments do not communicate because each one has to concentrate on the job it is given. And the company picks up the bill for this way of thinking. Cohabits are all about removing these silos by giving every coworker a bigger picture of the workspace they all occupy.

7.4.2 Cooperation

1. Cooperation does not mean others fitting in with what you are trying to do. That is a me-habit. These others are not there to fit in with your agenda, your work schedule. They are there so that the value of what they do is enhanced by what you do. Cooperation for the coexister means doing things that make the work of others more valuable for them and more rewarding for you.
2. Cooperating does not mean others working to make your life easier. That is a me-habit. After all, there is another life, not the afterlife but the new life, the coexisting life you recognize through cohabits. Cooperate means operate together, the two (or more) of you as one. Operations are individual, interdependent, and integrated. Now when you cooperate, according to cohabits, something new emerges: work that is the product of the two (or more) agendas, emerging from the one of you (plural), the coexisting you.
3. Every time a company makes a choice involving the development of a product, process, or supply chain relationship, that company is going to alter its set of capabilities. And its changing set of capabilities will affect the wherewithal of potential suppliers, and even alter the balance within the wider industrial context. Cohabiters will search out suitable means for scoping the landscape and figuring the dynamics of outsourcing decisions.
4. A company that outsources allows the supplier to develop its capability such that the company is even more inclined to work with this supplier in the future. Likewise, in-house work improves internal capabilities to a state where outsourcing becomes less of an issue. A cohabit to remember is that apparently straightforward commercial decisions can have sizable and far-reaching consequences.
5. In most cases of commercial alliance the extended enterprise lies dormant. An extended enterprise is, by its very nature, full of connections between the distinct legal entities. But so long as the maintenance and improvement of these links lie neglected, the true power of the relations that form an extended enterprise remains merely latent. The duty of the coworker is to activate these connections.
6. We do not yet find that project plans are properly integrated within an extended enterprise. The distinct entities are not even sure of each other's precise actions, focusing more on final delivery. Thus, the prime contractor, as the party responsible for managing the portfolio of collaborating companies, will draw together some sort of enterprise-wide plan, but this will not be integrated at the activity level. The cohabits solution lies in the generation of processes rather than plans.
7. When regarding an extended enterprise to which it might (or does) belong, a cohabits company asks: Are we a fit company to trade in this network? Do we have respect for what the network wants? Are we

willing to give ourselves to improving the ability of others in the network to make it healthier and in turn a better trading system? It is a *give* rather than a *get* attitude.
 8. When a company resolves to *inform* rather than *blind* customers, it is treating them respectfully and maturely, an attitude that will duly evoke a supportive response and with it a reinvigorated relationship.
 9. Cohabits shatter the notion that companies have authority or power over suppliers. A company has no such power unless the supplier itself had granted it. Sadly, many suppliers fawn in attempts to curry favor and win contracts from customers. In doing so they risk the danger of becoming subservient and resentfully so, engendering desires for retribution and blinding and gouging attitudes that help spread the cancer.
10. There has never been a more crucial time for companies to disciple others and, in so doing, make these disciples disciplers of others. That is the chain imperative for today's corporations, living in the extended enterprise era. The cohabits company will be motivated toward suppliers to make them better at what they do so they can give a better service, feel better about themselves, and want to treat their suppliers in the same way. Corporate discipleship is making disciples of companies capable of making disciples; it is the means for corporate success in the extended enterprise era.

7.4.3 Coeducation

 1. Preparation is key to successful accomplishment. Leave it out and failure is included in its place. What you save is buried beneath a mountain of costs. Finding a way back for preparation is a duty of every project coworker.
 2. If the project is very large, the preparation phase can be a project in its own right. In many projects, planning fails to be logical. But logic in planning always will and always does pay off.
 3. Planning that involves teams of people (coworkers) should *involve* each and every individual in those teams.
 4. Every company is trying to beat the clock, and whether they do is a major determinant of their success in the marketplace. That is tough enough. However, it gets worse. The clock is speeding up. Yesterday's heroics are never good enough for tomorrow's games, and the smart companies have clocked this relentless compression in time-space. What is their response? Make planning a core competence. Hold that! Make coplanning a core competence.
 5. There is no hiding place for anyone anymore for these two basic reasons: the clock is speeding up, and it is extended enterprises, not just companies, that have to beat the clock and get the job done. The habit for companies must be that they win by being a member of a winning team, not by going it alone. And a team that wins is picky about who

joins it. The qualification is be a good team player, and that means practice cohabits.
6. Change has to be arrived at consensually; after all, if any one party thinks of ways of speeding up the plan by slickening up its process, the knock-on effects of that change will be felt elsewhere and could conceivably slow down the plan.
7. The customer's process and the supplier's process are inextricably linked, and therefore, for either not to know what the other's processes are will have a direct effect on plans and therefore on timescales. In coplanning, the other side of the coin to Collaborative Product Development (CPD), it behooves the collaborators to let each other know what they do so that all of this doing can be synchronized for the sake of meeting the deadline placed upon the royal *we*.
8. Your company has to survive, and it needs to do this regardless of what the environment throws at it. But before it encounters that environment, it has a wider system to account for, and in principle that wider system has to survive the environment, not your company.
9. If you are paying attention to the wider system, having regard for it, showing it respect, then it is less likely to deselect you, more likely to consult you as regards responses to its environment, and indeed make you all the more influential within itself.
10. The true web is the web of people acting with the knowledge of physics and of programmatics to accomplish materiel flow and to enact, adaptively at that, the process architecture.

7.5 Time to Think

1. Below we provide four extracts, varying viewpoints on the problem of illegal immigration in the United States. They are taken from the BBC Web site.[6]

> Illegal immigration also has a negative effect on the U.S.'s economic security. It doesn't take a degree in economics to realize that a massive flow of low-skilled labor puts downward pressure on the wages of native-born Americans. These low-wage workers—who are largely paid off the books and without benefits—meanwhile cost the American taxpayer in terms of social services. Illegal aliens rarely pay taxes, yet they send their children to our schools free of charge, they receive welfare benefits, and they get free medical treatment. The best solution to our illegal immigration problem is to begin enforcing our laws. That means the federal government needs to get serious about prosecuting employers who lure illegal aliens into the U.S. with jobs. The threat of hefty fines and possible jail time will chasten employers' desire to hire cheap, illegal workers. We also need to

recommit to guarding our borders with more personnel, more technology and more money for physical infrastructure. And, we need to enable local police departments to aid the federal government in finding and deporting illegal aliens.

—Congressman Tom Tancredo

There is a tremendous amount of hypocrisy surrounding the debate. So many businesses are doing well on the back of undocumented workers—from the oranges that are picked in Florida to the tomatoes harvested in Illinois.

Yet, their basic rights, such as the right to a safe workplace and fair treatment, are not protected. Undocumented workers never file complaints for injuries sustained at work for fear of being sacked. Rich families in Los Angeles employ undocumented nannies to look after their children. They also employ undocumented housekeepers, cleaners and gardeners—many of whom have keys to their houses. How can we be called criminals when we hold the keys to the houses of some of the richest people in the state?

—Felipe Aguirre, deputy mayor of "sanctuary" town Maywood

With a wink and a nod, the United States government essentially allowed millions of people into the country to be employed in vital strategic industries.

These workers produce value and that value is appropriated by business owners. The worker is never remunerated fairly for the value he creates and the immigrant worker creates a value far and above what a native worker creates because he works for a lower wage, does not have paid holidays, a pension plan or sick pay. They make an incredible economic contribution to the economy. A fair exchange would be a streamlined procedure allowing them to legalize their status. The current legislation being debated by the Senate, the Hagel-Martinez compromise, would not be satisfactory for the immigrant communities.

—Nativo Lopez, national president of Mexican-American Political Association

President Bush, the tough-talking cowboy leader of the free world, will deploy 6,000 unarmed National Guard in "support" of U.S. Border Patrol along America's southern frontier with Mexico. This ostensibly will address the threat posed by daily infiltration of American territory by waves of illegal migrants and massive loads of contraband that include billions of dollars of drugs, weapons and exploited human beings. Our National Guard, armed with construction equipment and paper clips, will be placed in the midst of a virtual war zone to build roads and shuffle paper for an outgunned and undermanned federal Border Patrol. President Bush seems to suffer from the notion that the greatest problem facing border agents is insufficient office help, not corrupt Mexican military forces colluding with violent drug cartels and shooting at our people. The Mexican government also receives the benefit of $50bn sent home annually by illegal aliens.

—**Chris Simcox, founder and president of the Minuteman Civil Defense Corps, a volunteer group monitoring U.S. borders for illegal immigrants**

Create a systemigram that portrays the various issues alluded to in these four viewpoints showing how they interact, conflict, and possibly produce counterintuitive behaviors, thereby revealing the complexities of the situation and hopefully an enlightened problem definition, which we suggest is an essential prior to the formulation of executive action. See Figure 7.4 for an answer.

2. Using your understanding of Section 7.3 and any other relevant resources at your disposal, produce a storyboard of the C2 Constellation systemigram (Figure 7.3). How would you validate this? How might you defend the logic of this against critical attack that it was either incorrect or non-value adding?
3. The following structured text was developed from a review of the writings found within several open-source publications.[7] It constitutes an integrated rich text formulation of the particular views of Network Enabled Capability (NEC) strategy embodied in these papers:

UK defense policy directs the development of future UK operational concepts for land, air, space, maritime, and logistics. These operational concepts are guided by the UK Joint Vision, Joint High-Level Operational Concept, Effects-Based Operations Concept, and Defence White Paper. The operational concepts are informed by emerging concepts such as NEC to achieve the overall UK defense aim: "to deliver security for the people of the United Kingdom [customers] and the Overseas Territories by defending them, including against terrorism; and to act as a force for good by strengthening international peace and stability." NEC enables a flexible acquisition strategy to establish coherent acquisition programs (environment), these programs adopting an incremental approach to realize rapid technology insertion to achieve a

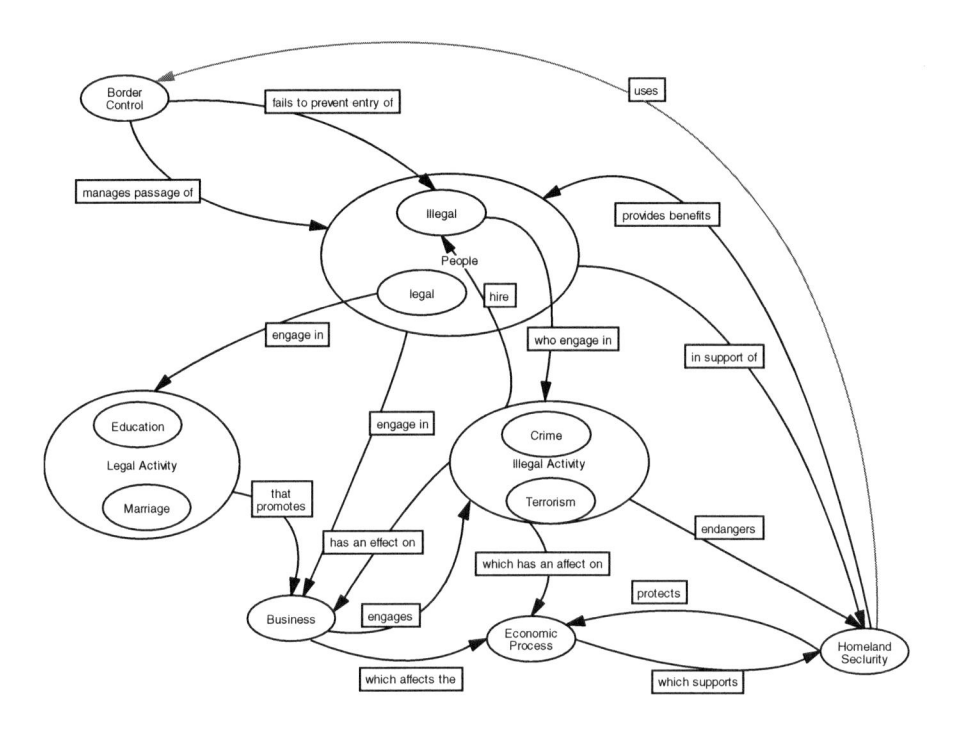

Figure 7.4 Illegal immigration.

net-ready force that exploits a network infrastructure to enable shared awareness. Shared awareness underpins flexible working to deliver synchronized effects that address the dynamic mission that is undertaking the defined UK military tasks to achieve military (actors) and nonmilitary effects (transformation). NEC requires an information infrastructure to provide secure and assured information access to support the network infrastructure and facilitate shared awareness. NEC also employs effects-based planning across government (owner), which requires a dynamic planning system supported by distributed tools and models to manage agile mission groups, thus enabling flexible working. NEC improves equipment integration of weapon systems, intelligence, surveillance, target acquisition and reconnaissance systems, and command and control nodes to facilitate the agile mission groups. NEC also enables networked support across public and industry to sustain agile mission groups used to enable flexible working.

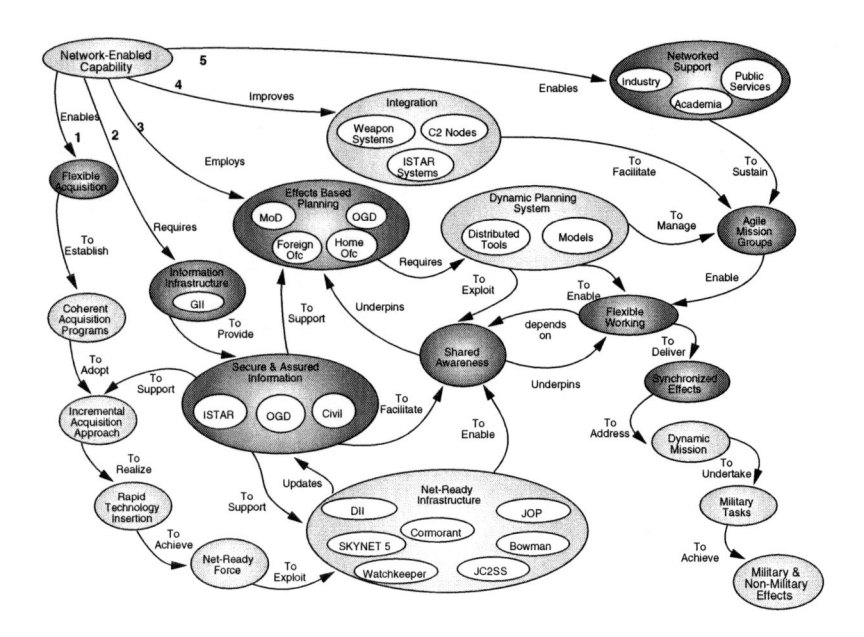

Figure 7.5 Net-centricity.

Consult this text to construct a systemigram, and use this to create a set of scenes to tell a story that you believe will validate your understanding and facilitate management support for the process of realizing the vision contained in the publications. See Figure 7.5 for an answer.

4. Tragedy struck at Virginia Tech on April 17, 2007, when a lone gunman, a senior undergraduate student, murdered thirty-two people and wounded twenty-five others before killing himself. It was a graphic reminder of the phenomenon of school shootings prominent events, among which are the slayings at the University of Texas, Austin in 1966, and at California State University, Fullerton in 1976; the Stockton massacre in California in 1989; and the Columbine High School massacre in Colorado in 1999. The phenomenon is sadly very much a twenty-first-century problem with apparently little that has been learned about prevention and remedy having been effectively executed. Many are involved in both the problem and its feasible solution. In the school system, viewpoints are valid from teachers, students, administrators, and counselors. On the home front, parents, siblings, friends, and the community at large must be heard. These crimes, perpetrated by "unstable kids," not only are aided at the scene by guns and bombs but also are previously equipped via neglect, violent materials, family instabilities, and an acute sense of being outcast. Obvious school safeguards include security cameras, security guards, locked entry points, and

metal detectors. At a higher level, prevention strategies might include gun control laws, character education, and social programs. Integrating these elements together is nontrivial, and many more perspectives and systems elements can be added. Using whatever resources you can draw upon, create a defining statement of this phenomenon and portray it as a systemigram, with suitable storyline, that captures the essence of the problem and offers multiple stakeholders, for their consideration, a well-thought-through approach to its eventual eradication.

Endnotes

1. Picciotto, R., and D. Paisner, *Last Man Down: A New York City Fire Chief and the Collapse of the World*, Berkley, New York, 2003.
2. Ames (CIA analyst) and Hanssen (FBI agent) sold U.S. secrets to the Russians. See, for example, http://www.usnews.com/usnews/culture/articles/030127/27traitors.hanssenames.htm.
3. See www.intelligence.gov.
4. See http://www.wmd.gov/report/report.html.
5. See http://www.af.mil/library/airforcepolicy2/january/april.asp.
6. See http://news.bbc.co.uk/2/hi/americas/4989248.stm#middle.
7. The Defence White Paper: Delivering Security in a Changing World, UK MoD, 2003; Joint Doctrine Note 1/05, UK MoD, 2005a; JSP 777, UK MoD, 2005b; and Ministry of Defence, United Kingdom.

chapter eight

Of

We have come to think of an automobile in terms of the assemblies from which it is made. Chief among these are the engine, the transmission, the suspension, the steering, and the interior, which of course includes seats, controls, and instrumentation. Holding these together is the chassis, a primary assembly all of its own, and the body of the car. Now for some this last item is the most important. Appearance is everything. It denotes status and can convey an appeal to others that is a definite attraction the car's owner may well have in mind at the time of purchase. Other owners may take different perspectives emphasizing ride quality, or safety in the unlikely event of a serious collision, or the infotainment quality the vehicle carries in terms of satellite navigation and radio (maybe even video). Regardless of the priority that drivers give to the qualities of the automobile, it is almost natural to think of this system in terms of its primary elements, any one of which can undoubtedly be broken down even further into smaller pieces. The icon that stands for this decomposition is that of hierarchy, shown in Figure 8.1. In Chapter 2 we noted this form of organization to be a powerful force that grants stability to the various components, making it easier to organize these even further as we move from basic pieces to the automobile itself. From bottom to top, hierarchy is an unrivaled mechanism for bringing order. It is a significant *of*.

When Moses had the awesome privilege of leading his people Israel to the promised land, having witnessed amazing miracles of God in their deliverance from Egyptian slavery, he was advised by his father-in-law to use this particular *of*. Let us say that the population was a million. Moses was the leader acknowledged by all. It was he, with God's help, who had set the people free. And whenever anyone of them had a problem he or she came to Moses. One problem for each one amounts to a whole lot of problems for Moses. And it was beginning to tell on him. Jethro saw this and wondered why Moses should be the only judge with a long line of people at his door, day after day, waiting for arbitration, advice, judgment, and leadership. This was not good. So Jethro told Moses to find capable men of overseeing thousands, hundreds, fifties, and tens. The qualities that made these people a judge, standing alongside Moses, were that they feared God, were trustworthy, and hated dishonest gain. The tough cases still came to Moses, but all the rest went to his fellow judges, who shared the load and made life a whole lot easier for Moses. Hierarchy works. It is a great *of*. But it is not the only *of*.

In this chapter we want to share a different *of*, an alternative organizing paradigm, and compare this with the well-known hierarchy *of*. We also want

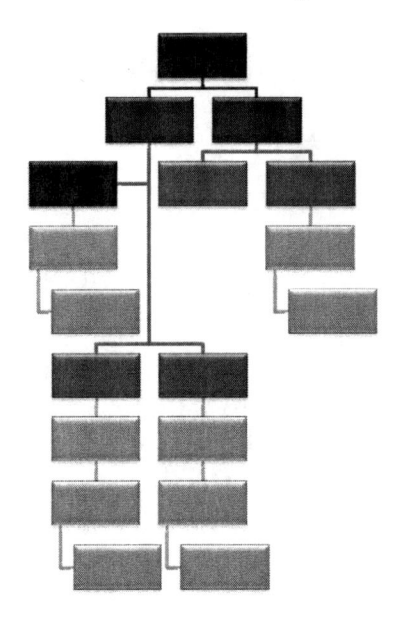

Figure 8.1 Hierarchy.

to look at those kinds of situations that suit one better than the other and provide guidance to thinkers and practitioners as to when, where, and how to use one in preference to the other. Of course, use presupposes design intent as opposed to naturally occurring. In that case we want to observe why the hierarchical *of* occurs naturally in the way that it does and works well, and when the other *of* seems to work better. Our other *of* is defined by its own icon (see Figure 8.2), and if this appears to be somewhat reminiscent of a network, with nodes and links, hubs, spokes, and clusters, for the sake of convenience that is what we will call this *of*: the network *of*.

8.1 Social Networks

We know that it was a good thing for Moses that he took Jethro's advice. And it was a good thing for the people also. They did not have to wait in line so much, and if they still needed their case to be heard by the head honcho, they would have got to him far more quickly than had they had to take their turn with everyone else. In this respect, that of getting problems resolved, hierarchy works. But we also know that this form of organization does not dictate everything in the lives of individuals and in their ways of socializing and pursuits for social togetherness. The boundaries that are constructed to form groups of ten, fifty, hundred, and thousand work for matters of judgment but not necessarily for everything else. People congregate together for all kinds of reasons. Togetherness has its own forms of expression and its own way of organizing people. Take, for example, one of our first loves, movies.

In the movie industry there are clearly stars; in fact, there is an acknowledged A list, and these include the likes of Tom Hanks, Tom Cruise, Jack Nicholson, Robert de Niro, Al Pacino, Julia Roberts, and not many more. For every name on the A list there has to be a thousand wannabes. But this does not mean a clear separation of the two categories. If a wannabe makes a movie with an A list star an association is formed. It may not be lasting or impressionable, but it could be. The wannabes know it is a big break to appear in a movie with an A list star, and the A list star knows the next person to join that list will come from the wannabes category. They did. And it is in their own interests to keep a lookout for rising stars, who might be able to help them as they gracefully slide down the ranks and before they

Chapter eight: Of 145

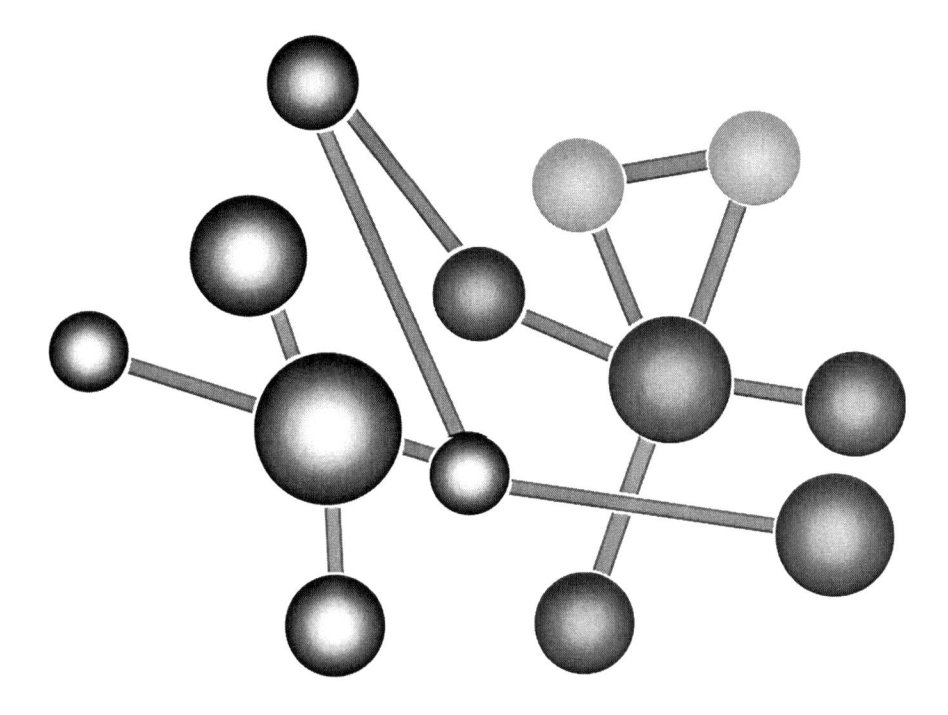

Figure 8.2 Network.

eventually disappear mournfully from public view. Movies make for associations, if not life sacrificing relationships, and in that sense they produce a form of organization, one that is essentially of a network variety: a social network.

One actor who has gained a reputation for social networking, in the sense of appearing with a very large number of different actors, is Kevin Bacon. Such is this guy's fame for having an eclectic collection of associations, via the movies he has made with other actors, that all other actors can be said to have a Kevin Bacon number. This is 1 if you have appeared in the same movie as Kevin and 2 if you have not appeared in the same movie but have appeared with someone who has. Your Kevin Bacon number is 3 if your limited claim to fame is to have appeared in a movie with someone whose Kevin Bacon number is 2. And so on. It turns out that less than a hundred actors, out of a total of almost a million, have a Kevin Bacon number greater than 6. The average Kevin Bacon number is less than 3. See Figure 8.3 for a depiction of this phenomenon. This type of outcome is termed small world, and is the subject of the final chapter.

With this kind of organization almost no one in the nation of Israel would be more than six handshakes away from Moses, with a high probability of being less than three handshakes away. In a hierarchical organization almost

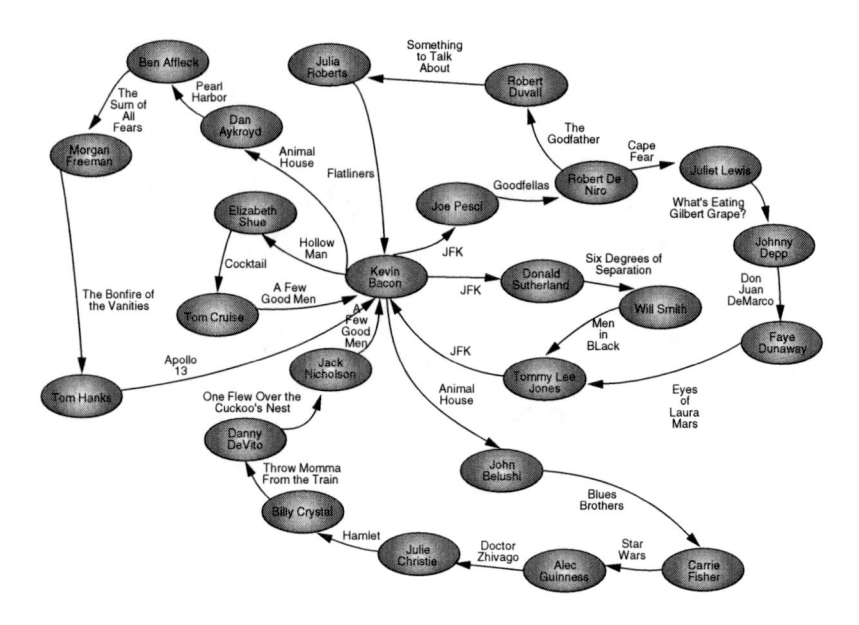

Figure 8.3 Bacon social network.

everyone is exactly six handshakes away from Moses. One thing this particular *of* does, in principle, is bring people closer together, which is unsurprising since the associations are formed by the people themselves as opposed to an external rule that is imposed upon them. But then closer together is not always a good thing!

8.2 Order Forms

What we are interested in knowing is what are the essential characteristics to any particular form of organization, to a specific *of*, be this the hierarchy kind, the network kind, or some other. If we say that a system is a collection *of* parts and relationships gathered together to form a new whole with new attributes, properties, functions, and capabilities, then what we are asking is this simple question: What is the nature of the *of* in "collection of?" Put another way, how is this gathering together made? Who makes this togetherness what it is? Who says what the *of* is and why should this *of* work well, or not, for any given kind of system?

With hierarchy order appears to be imposed. Now this imposition may have the consent of those on whom it is imposed. Moses was a great leader who loved his people and wanted them to know God in the way that he himself knew him. His rule would assuredly be benevolent and the people trusted him—after all, he had gained them their freedom and was leading them to a land flowing with milk and honey. They could be confident that hierarchy would be in their best interest because they knew that was what Moses wanted for them.

By contrast, order appears to emerge with the network form of organization. Individuals make choices, perhaps somewhat limited if you are a wannabe, to appear in the same movie or not and thereby form an association of indeterminable depth, at least a priori, but valid nonetheless. Organization appears to form on the basis of consenting adults who may act in the interest of others, but not necessarily. The collection of parts adds a collection of relationships forming a whole that is more than the sum of a series of bilateral associations. Some parts are undoubtedly more attractive than others, but that attractiveness is influenced by the reception the movies get and the reception wannabes get when they are put in front of the audiences. There is a dynamic to this organization that defies the very notion of a priori goodness and therefore orderly imposition. That is what we have so far based on a tiny sampling of social networks.

The essence of the network *of* appears to rest on autonomy of the parts; willingness of these parts to belong to the society (actors in the movie industry, say); connections that form in a rather ad hoc manner, at least as far as the society is concerned, slightly less so as far as two fellow actors may be concerned; and an emphasis on emergence as opposed to a priori determination of form. Of course this paradigm does not rule out an emerging hierarchy; after all, isn't this what the A list represents—a sort of ruling aristocracy? But an emergent hierarchy would be a special and possibly mystifying form of organization with various network architectures being the obvious outcome.

The guiding rule for the hierarchy *of* is span of care. (Additionally there may be criteria for the selection of leaders.) After that it is up to the various pieces of the hierarchy to remain stable, a property that itself will be governed by the efficacy of the hierarchy to serve the interests of the individual parts, which in the end is the overarching rule for all forms of organization. The guiding rules for the network *of* are exercise your individual liberty, understand the significance of society and take your place in it whatever that is, be as influential as you can in forming relationships for yourself, and watch what emerges because whatever this is will affect your liberty, the society and your being part of it, and the relationships you keep, break, and adopt.

8.3 Technology Networks

At this stage in our discussion it appears that the hierarchy *of* can apply to technology systems, such as an automobile, and to people systems, such as a corporate organization. The network *of* clearly applies to people systems, but it is not clear if or how it applies to technology systems, notwithstanding the clamor for network-centricity from industry and commerce generally, and for net-centric operations/warfare from the Department of Defense (DoD) in particular. Of course we can build networks using technology;

transportation and communication networks have long since been part of the engineer's repertoire, with highways, railroads, airways, and telecommunication systems being prime examples. That much is clear. But can we build pieces of technology, components of systems or systems themselves, in accordance with the network *of*, as opposed to the hierarchy *of*? And what might this mean? How would such pieces of technology differ? In both design (form) and capability (function)? And what methods (fit) do we use? The same ones for both or different ones?

From its inception, or thereabouts, Sun Microsystems gave us "The network is the computer." It sounded great. No one really knew what it meant. It turns out Sun Microsystems did not know what it meant. But not only did it sound great, it seemed right. The history of computing had focused on the machine, the artifact, the thing that processed data. It had inner workings; it had an input device, so that you could tell it what to do and with what information to work; and it had an output device so you could check the results of a processing operation. We had a strongly node-centric view of the computer. It was a machine. It may have been made up of much complex circuitry, a labyrinth of wires that suggested its interior was a network, but it, the machine, was not a network. It was discrete, monolithic, integrated, and mechanistic. If it was anything, it was a hierarchy of elements: a central processing unit (CPU), input/output (I/O) devices, memory elements of various forms, and a clock, to keep its heart beating. Whatever was the network was "out there," outside of the machine, something that the machine—the computer—would get plugged into. A network of many machines, and a variety of devices, spread far and wide. Now we were being told that this network was the computer. It was like being told that the particle was really a wave. The computer is no longer that which is in our possession, in our grasp, held by our hands and under our control, but it was really part of an ever-rolling ocean that throws up in ways beyond our control a perpetual series of waves. Perhaps the mantra was pointing to new forms of architecture in which computers would operate? The client-server architecture certainly shattered the traditional notion of the mainframe and its smaller, more agile progeny, the minicomputer. Suddenly you were not alone; you were part of a computer society. You still had your autonomy, but if you were smart you joined this society. Your belonging benefited you and affected it, stimulating its attractiveness and growth, and expanding both its connectivity and diversity, with each new client bringing a unique individuality. Unsurprisingly, this architecture became the forerunner to a global technology system we know today as the Internet, overlaying which is the World Wide Web, the global village's new wheel. Maybe that is what Sun Microsystems meant? Who knows? It interests us that one of the founders of that eminent corporation left a $100,000 check with two young men who had recently dropped out from Stanford's computer science Ph.D. program in order to found their new corporation. The university's Office of Technology Licensing had failed to sell their invention, a search algorithm, including an unsuccessful approach to Alta Vista, who at the time passed on buying it for $1 million, maybe because at the time its corporate owners, DEC, were fully preoccupied with survival, themselves being bought out by Hewlett Packard.

That pass on $1 million was a really bad call. Andy Bechtolsheim's initial investment is today estimated to be worth $500 million and the two dropout kids, Larry Page and Sergey Brin, are multibillionaires. Their company Google is rapidly approaching the size of Microsoft, and its core technology is all about finding the right information, for maybe as many as 10 million people every day, from the network that is the computer. How the wheel turns!

Both the Internet and the Web are nonhierarchical. True, you will find aristocracies present, because these global technology systems conform to the same small world architecture as movie actors. But it would be a mistake to take this emergent feature and from it argue that these technology systems were hierarchical or had been designed thus, that is, the a priori design intent had been hierarchical, in the manner of a simple automobile. So it is possible to think of technology systems as having an organization that mirrors the network *of* rather than the hierarchy *of*. But can the same be said of a mere automobile, and what would this mean? For the automobile itself and for its designers?

8.4   Less Auto, More Mobile

One of our favorite movies is *Planes, Trains and Automobiles*. Two guys, Neal Page (played by Steve Martin) and Del Griffith (John Candy), are thrown together against their will and beyond their control two days before Thanksgiving. Neal is a well-to-do marketing guru living high off the hog in a swish Chicago suburb. Del is a shower curtain ring salesman whose home is his cavernous valise decorated with the treasure of endless travel. Neal, needless to say, is not attracted to Del, so sleeping with him in a shabby Wichita motel marks the nadir of an unwanted relationship. Not so. Headed back to Chicago in a rental car both are warned by motorists headed in their direction but on the opposite side of the freeway that "You're going the wrong way!" How would they know? How would they know which way Neal and Del are headed?

The hapless duo discover the truth as they speed sandwiched between two giant semis headed in the opposite direction, going the right way. Some nadirs must be local minima!

When we design an automobile are we headed the wrong way? We remember our first car. You had to crank it. We were blessed to have a heater. Radio? Forget it. A spare fan belt was essential equipment. As far as we are concerned, automobile design has definitely been going the right way. And yet, maybe there is another way of looking at this?

The need to regard an automobile hierarchically is seemingly inescapable. A car must have an engine, a transmission, a suspension, and so on. Each one of these assemblies in order to be what it is must have an internal structure. We could and we do go on down the line, each branch forming part of an inevitable irresistible hierarchical structure. The *of* of the automobile is clearly laid out for us. Each part does its work, and in so doing the car has propulsion, it has traction, it is navigable; the vehicle can be made to move forward, accelerate, brake, and stop. It can be made to turn. It can get you where you want to go. In comfort and safety, all the while with you being informed and entertained. The car is built with all these functions so that the joy of driving is maintained. So what is the problem? No problem at all if this satisfies. But in the increasingly sophisticated world in which we live we demand more. We want no traffic jams. We want our car to be our office and a place where we can plan and organize our leisure. We want no collisions, no wrong turns, and no surprises. The burden of all this demand falls on our shoulders because what engine can help us book our theater tickets, what suspension system can avoid traffic jams, what transmission system can eliminate collisions, and what navigation system can serve our office needs? We have chosen this mismatch of functional assemblies and driver requirements deliberately. Our motive is to shatter the notion that specific components can deliver on specific functionalities. Only the collection of them, the network of them, can do this. And our networked automobile can deliver on each and all of them and more as yet unspecified and unmet needs. How?

In order to avoid a serious collision, many things come into play: speed, braking, weather conditions, traffic conditions, driver's state of mind, steering, information overloading, tire pressures, and vehicle dynamics. This is just scratching the surface. And all of these variables, snapshots of the pieces of technology of the car and the psychology of the driver, interact continuously and probably in nonlinear fashion. Driver skill smoothes out the nonlinear wrinkles. Most of the time. And this skill should never be abandoned or eliminated. But neither can it be relied upon entirely. And in the blink of an eye maybe it should not be relied upon at all. In the blink of an eye, the nervous system of the automobile could make informed decisions, if it is allowed to and if the society of technology pieces from which the car is made is appropriately consultative, fully informative, and given choices to act in their own best interests and always overridingly for the good of the car's occupants.

Think of the car now in this way. Its parts, every single one of them, of which there could be a total of over a million, have two basic components: an inner core and an outer core. The job of the inner core is to keep the autonomics of the part—the thing that the part does entirely well and for which no oversight or management attention is needed by any other part. The core knows what to do and does it slavishly, autonomically, perfectly, promptly, and optimally. The outer core tells anybody who is interested who the part is and how available it is. It also wants to know what is going on, what is expected of it, and how serious the "case" is. This is the part's master and at the same time, on behalf of the part, a servant in the technology society in which the part is included. In this way the part has autonomy, but it also has every opportunity to be part of the society—autonomously. It has the opportunity to belong not merely as a slave but as a contributing member of society. It has an openness to connections with other parts, more precisely the outer core of the other parts since the only thing that commands a part's inner core is the part's outer core.

By this design, the network *of*, the major parts and maybe down to much lower levels have an openness with one another that forms a stronger community of technology pieces, thereby enabling enhanced cooperation via shared understanding of whatever demand is placed upon them. The drive is no longer the sole responsibility of the driver, whose previous slaves were summoned by his crude signals to speed up, slow down, turn, and shut down. Now the drive is the collective responsibility. It always was but in the most unintelligent way possible. Now we have the wisdom of crowds to leverage, a collective wisdom built upon autonomic expertise and autonomous networking.[1]

The luxury car makers are already headed this way. They are going the right way. In time the trickle-down economics will bring such benefits to the lower end of the automobile spectrum. But when we all come to realize the actual costs to our society of missed opportunity costs and needless road deaths, maybe then the trickle will turn to a flood and our use of the hierarchy *of* will be supplemented and complemented by the network *of*. Maybe.

8.5 *The Price and Prize of Togetherness*

Revisiting what we have said thus far, we are asserting that the word *of* is significant. It has more than one meaning and we have mentioned two. Both these meanings are interpretations of that far too little valued part of the definition of the term *system* "gathered together." What *of* represents for us is a key to making sense of togetherness. How things (we are not ashamed to use such an imprecise term) come together, stay together, reorganize their togetherness in the light of success and failure, embrace new things and new forms of togetherness, and—if it is possible—communicate this success (and failure) to others, in the present and the future, is the essence of understanding what a system is and how this understanding can be applied to discoveries in biology, inventions in technology, and comprehension of society.

The two meanings of *of* we have articulated are captured by the terms *hierarchy* and *network*. The former is strongly exhibited in nature and much prized in technology development and societal structures. It has pros and cons. The pros are stability, robustness, and clarity of order. The cons are rigidity, pedantry, and intolerance toward insubordination. The latter too has pros and cons. The pros of network are flexibility, agility, and emergence, by which we mean the observance of patterns over the longer term rather than the short term. This can be an advantage to those having to deal with security issues. The cons are summed up in one word: messiness. The lack of a priori order and not knowing what order will emerge and when is debilitating for some. The lack of clear lines of communication, reporting, accounting, and jurisdiction is also a concern. The challenge to being both independent (fending for yourself) and interdependent (finding neighbors with whom you can better succeed) is a stretch for many.

Knowing these distinctions can be useful when a choice of one or the other presents itself. What we are interested in is suggesting that both apply simultaneously, and in fact, they are merely different perspectives of the same objective reality. It is this insight that most usefully informs our guidance as to which *of* applies when, how, and to what. We shall explain.

Imagine a world littered with myriad identical objects. There are lots of them and they all look the same. And they are not connected. Over time these objects draw close to one another, reducing the number of objects, because discrete combinations are formed, but adding to the connections between objects. An intriguing behavior is exhibited, one that was not in evidence when the world had lots of identical objects that had no connections. Fascinated as we would be with this new behavior and observing this, looking for patterns and explanations, we might be even more interested in the causes of the togetherness we witnessed. Instead of simply or solely delving into the nature of this behavior, we might ask: What was it that drew the objects together? What was it that made them form connections and new objects in the way that they did? And what caused the breakup? Was this failure or success? Is there an invisible hand guiding all of this or are these objects all that there is? Finally, how would we find answers to these and all other questions?

Well, if the objects are all that there are, it is futile, at least in the first instance, to go looking for the invisible hand. And there is not much to observe other than the objects, unless we observe their separation. And we should be prepared to see in the separation a source of explanation for the eventual togetherness. After all, it may have been a very long time before any sign of coming together was observed, a time long enough to give up and conclude that nothing ever happened.

Let us picture these identical objects as spheres. All spheres are identical in this respect: they all have an outer coating and an inner core. In that sense they all remain identical. However, the outer coatings are all different in terms of density and substance. Some outer coatings are very thick and rich in material (the exact composition of which still remains undetermined), others are rather thin and uninteresting. The inner cores similarly differ. These geometrical and substantive differences, we hypothesize or conclude, lead to the drawing together and beyond that whatever emerges in structure and dynamic.

With no scientific apparatus whatever to determine the nature of material in the outer coating or the inner core we are left only with abstract thought to guide us in the search for explanations for the endless cycle of togetherness, further separation and dynamic behavior of both structure and functionality.

(Maybe this includes giving off light, sounds, smells, and other such sensory experiences that inform, educate, and entertain us.)

What we believe is that the outer coating is at least partly responsible for the drawing together and the separating apart. But the outer coating works in utter fidelity with the inner core. They at least are a team. They never separate for any given object. They change themselves but they never leave one another. It is as if the two have a continual dialogue as to what to do as an object. A single monolithic being with just these two elements, coating and core. The coating is responsible for negotiating (with other objects via their coatings). Some of the subject of these negotiations is what the core can do, without any burden whatsoever being placed on anyone else. What these cores can do, no other core can do. And that is what attracts other objects. They want some of that, and they can have it by coming together. And apparently that is the only price to be paid, being together, since the core will take no energy or attention from anywhere else to do what it can do. However, negotiations cost. But that is what they are for, if in fact communications between a coating and a core are judged by the two to be unfulfilling. "Let's see what others have to offer," says any given object, working as a team, core and coating. Maybe something happens. Maybe nothing. Whatever. At least now we have a sense of dynamic. Objects talking to themselves and then possibly to others and on the basis of these communications doing something about it, that is, coming together.

At this point in our highly abstract conception we attribute competence to the core (core competence) and communication to the coating (which is not to say that the coating is incompetent). The rationale for the togetherness is to draw benefit from the competence of others and to see what emerges, as well as an exhibition by a new "combined" object of not only the two (or more) core competences, but we may discover new competences that may not have been foreseen. A case of what happens if? The notion of communication can be further articulated by the notions of belonging and connectivity. These address questions like: Do I want to partner with these others? Do I want to belong to what we together form? What kinds of connectivity must I make? To these two we must add diversity. Why? Because the competences are different and the competences combined may well produce something different. Plus the combination calls for agreements about belonging (partnership) and connectivity. Maybe objects have to be prepared to change a little in order to make the connection and achieve the belonging? What remains sacrosanct throughout all this is the unbreakable, unshakeable partnership between coating and core for any given object. The coating is the shield, the filter, through which agreements, connections, and belongings must pass. It is also the instrument that triggers the competence into action. Nothing else can do this, though of course the coating may now have to take account of the connections and the belonging factors when it gives the signal to the core to do its thing. So unquestionably throughout this combination phenomenon, objects remain autonomous, characterized by the autonomics of the core when it does its thing. But this autonomy is certainly now influenced by the togetherness, the negotiations that leverage diversity, establish connectivity, and affirm belonging. See Figure 8.4 for a graphical portrayal of this conceptual chemistry.

From this otherwise uninteresting world of myriad identical objects, an interest that is uncovered by the existence of core and coating and an inquisitiveness that explores the notion "What if togetherness has value?", we elucidated five vital characteristics: autonomy, belonging, connectivity, diversity, and emergence. We necessarily add this fifth characteristic because the inquisitiveness (What happens if?) and the phenomena of new behaviors resulting from combination strongly suggest this. With these five characteristics, what sense can we make of systems thinking and this new kid on the block, the system of systems?

8.6 The System of Systems Debate

For the military the watchword in the system of systems (SoS) community is interoperability.[2] In global commerce it is partnership.[3] In our view these are equivalent terms. In U.S. military doctrine the notion of interoperability extends far beyond the conformance of complex technology systems into the need for alignment in many dimensions—cultural, command, cognition,

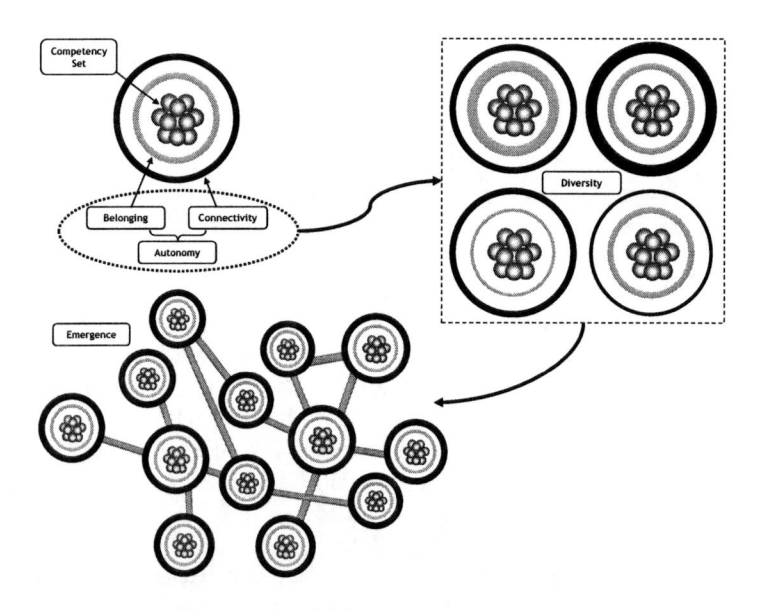

Figure 8.4 Biology of systems.

and country—raising the bar for the military services to the joint level and for countries to the multinational coalition level. The words may change and the specifics of the problematique certainly do but the game is the same: making disparate, diverse, autonomous, and asynchronized entities work together, without losing their individual sense of purpose and without loss of idiosyncratic capability, in order to realize some higher-level otherwise unattainable purpose. For systems people this has always been the challenge: getting it together.

In attempting to come to terms with the SoS phenomena the literature is largely agreed on a few crucial descriptive terms, most eloquently summarized in Sage and Cuppan:[4] operational and managerial independence, geographic distribution, emergent behavior, and evolutionary development. Others have defined descriptive terms such as enterprise activity,[5] networks of heterogeneous systems,[6] autonomous embedded systems,[7] social infrastructure,[8] and knowledge-based systems.[9] In our attempts to characterize a SoS we found few studies that attempted to transcend domains. Of these Bar-Yam[10] crossed three domains (i.e., biological, social, and military) and identified characteristics of a SoS as evolutionary development, emergent behavior, self-organization, adaptation, complex systems, individual specialization, and synergy.

While many have pursued a definition of SoS, like Sage and Cuppan and Bar-Yam we have pursued a characterization of SoS and asked ourselves this fundamental question: What characteristics can we posit that can not only help define a SoS but also distinguish it from a system that is not a SoS? By so doing, we might provide a set of continua for a systems typology showing the quanta that perhaps define stages in emergence between the two and along their continuum. We would want our chosen distinguishing characteristics to be traceable back to the center of gravity in the arguments as to what constitutes a SoS, as evidenced in the literature. This we have done,[11,12] and Table 8.1 summarizes that. But we want our distinguishing characteristics not only to reference the past but also to maintain continuity with our current philosophical thinking as articulated in the interdependent development of legacy assessment, capability envisioning, problematique demystification, and interoperability framework.[13] Finally, we want our distinguishing characteristics to have practical utility, to be usable and used by conceivers, developers, and managers of a SoS. In this way they refer to the past, are in line with current thinking, and can project into a future that is held by as yet unknown stakeholders.

8.7 Essential Characteristics

The five distinguishing characteristics we have chosen are autonomy, belonging, connectivity, diversity, and emergence. In this section we want to explain their meanings and origins. In the remainder of this chapter we will concentrate on their utility and development.

Table 8.1 Differentiating Characteristics

Characteristic	Definition	Motivation	Provenance
Autonomy	The ability to make independent choices; the right to pursue reasons for being and fulfilling purposes through behaviors.	Legacy systems are indispensable to a SoS; the SoS has a higher purpose than any of its constituent systems, independently or additively.	Managerial and operational independence.
Belonging	Happiness found in a secure relationship.	Legacy systems may need to undergo (radical) change in order to serve in a SoS.	Shared mission.
Connectivity	The ability of a system to link with other systems.	Legacy systems targeted for an envisioned SoS are very likely highly heterogeneous and unlikely to conform to a priori connectivity protocols; the SoS places a huge reliance on effective connectivity in dynamic theaters of operations.	Interdependence, distributed, networked, multiple solutions, interoperability.
Diversity	Noticeable heterogeneity; having distinct or unlike elements or qualities in a group; the variation of social and cultural identities among people existing together in an operational setting.	Legacy systems were most unlikely to have been purposed to work together prior to targeting the envisioning of the SoS; the SoS can only achieve its higher purpose(s) by leveraging the diversity of its constituent systems.	Independence, diversity, heterogeneous.
Emergence	The appearance of new properties in the course of development or evolution.	A boundary is indispensable to a system; all systems are emergent; emergence requires a well-defined boundary; a SoS has dynamic boundaries but always clearly defined; ergo, a SoS should be capable of developing an emergence culture with enhanced agility and adaptability.	Evolving, intelligence, synergy, dynamic, adaptive.

8.7.1   Autonomy

The reality of legacy systems relative to an envisioned SoS is inescapable just as individual freedom of choice is incontestable. For human beings autonomy is defined as a person's ability to make independent choices. What of a system? Each legacy system that is envisaged to become a constituent system in the SoS must be accorded autonomy, the right to pursue reasons for being and to fulfill purposes through behaviors. Respect for this autonomy is paramount, and it is a respect that the SoS itself must pay. That is not to argue that the legacy systems cannot be migrated or morphed to more aptly serve the SoS, but it is to argue that such transformation must be out of respect for that constituent system's autonomy. To do otherwise is to imperil that constituent system's functionality and essence of being, which might then be lost to the SoS, a foolish thing since it is these features that are wanted for inclusion. We argue that the capabilities of the SoS are enhanced by the exercising of constituent systems' autonomy, and that the opposite is true of a system that is not a SoS, whereby its parts must cede whatever autonomy they might have had in a totally subservient act of granting autonomy to the system.

Smuts[14] introduced the term *holon*, which later was explained in more detail by Koestler[15] as being that which is both whole *and* part. This term aptly fits constituent systems relative to a SoS. However, it is proposed that a SoS cannot be so called on the basis of structure alone, including hierarchies and holarchies. It must also qualify on the basis of dynamics, for which the remaining distinguishing characteristics provide further explanation.

8.7.2   Belonging

Just as legacy systems are a reality so also is the problematique that goes unsolved by these systems, singly and additively. By the same token the envisioned SoS is a reality if only in concept. Someone or some persons see that the SoS, by making use of the constituent systems via a new framework, will in a real sense deal with the problematique. So there are two new realities: the problematique and the envisioned SoS. This makes the second differentiating characteristic, belonging, a key one. The SoS cannot translate from conceptual reality into physical reality without the constituent systems belonging. But why should they? What is in it for them? Who can make them belong? What will become of them once they do belong, given that they will not lose autonomy? How will they belong? Will they continue to belong, come what may, or will their belonging be strictly conditional? Can they exit without hurting the SoS or themselves?

The parts of a system (that is, not a SoS) have no choice in the matter of belonging since they have no reason for existence and no dynamics to contribute without belonging. Parts in such a system are integral and the system cannot function without them. In a SoS the parts, also wholes and therefore holons, are integrable, that is, capable of being integrated. It is proposed that for a SoS there must be negotiation between it and each constituent system about the latter's belonging and the former's acceptance. There will be manifestations of the problematique when it is better for a constituent system to unbelong or for it to be believed that it does not belong when it actually does. We must continually bear in mind that the existence of the SoS is to confront a perpetual problematique for which no single point solution, no single system, is adequate. It is not about the system as such but about the SoS capabilities for resolving or addressing the problematique. Hence belonging becomes a core competence or stratagem available to the SoS for dealing with the problematique.

8.7.3 Connectivity

For the U.S. military, interoperability translates into net-centricity.[16] They want the same powers of connectivity among their warfighters, commanders, and others who "need to know" that global commerce has acquired via the Internet and the World Wide Web, instruments that have transformed business models. No surprises there, except there is an irony considering the DoD's chief concern is with an enemy that is organized as a network—testimony to the maxim "fight fire with fire?"[17] Later we will get into the practical application of the connectivity distinguishing characteristic, but for now we want to explain its central importance.

Most designed systems require the relationships between elements to be designed simultaneously with the design of the elements themselves. Thence connectivity between components is considered alongside the design of these components, regardless of the topology of the connections, be this integrated, distributed, hub and spoke, or whatever. This design pattern normally leads to hierarchies (or holarchies) and a valued stability in development whereby parts or subsystems are themselves stable, enabling a gradual buildup of the designed whole, which of course must also be stable. However, many such wholes or systems (that are not of the SoS kind) have designed connectivity to their environment, and this is fixed; it cannot emerge. The problematique that confronts a SoS will ensure that such limited, presciently designed connectivity leads to inevitable system failure.

Therefore we argue that a distinguishing feature of a SoS is that the internal connectivity of the SoS is not presciently designed but emerges as a property of present interactions among holons. Net-centricity is a form of prescient design, enabling full connectivity by supporting interactions and connections between all the elements, according to defined protocols. Further, it supports extension as more holons are added to the SoS, provided that these holons conform to the protocols. In our scheme for a SoS this connectivity is itself adapted as holons enter and exit the SoS. And this takes place in a way that enhances the connectivity or interactivity of the SoS with its environment, that is, dealing with the problematique. In the context of this discussion, connectivity has to do with a lot more than just topologies and protocols and interoperability standards, although it does address these practical matters, and is more concerned with the agility of structures for essential connectivity in the face of a dynamic problematique that defies prescience.

8.7.4 Diversity

Imagine soldiers who are not soldiers but who wage war that is not war. Citizens who are loyal to no nation-state to which they notionally belong, but who really belong to the vision of an integrated, faith-based, global-wide superpower governed by a single ruler headquartered in the Middle East. Imagine warriors who are not trained in their country of origin but in foreign lands, including that of their enemy, and trained by that enemy in skills needed for battle. Fighters who have no armor or weapons to speak of save the legacy systems of their enemy, namely, the Internet, cell phone technology, Boeing aircraft, the air transport infrastructure, up to a point, and box cutters. Can you imagine that? If we had, could 9/11 have been averted? Our problem in perceiving these threats to an extent lies in our inability to cope with diversity. Ross Ashby posited a law of requisite variety asserting that for a system to be sustained it must have at least the same number of degrees of freedom as the environment in which it operates. To paraphrase, interior diversity must match exterior diversity, or the boundary that separates them is futile. Post-9/11 efforts have largely concentrated on the boundary—understanding it, strengthening it, and in one sense extending it, for example, by military occupation of some nation-states. Greater attention is now being given to increasing interior diversity and reducing exterior diversity, a role, we argue, that falls to SoS thinking and acting.

Engineers have a problem with diversity, summarized in the maxim "keep it simple, stupid (KISS)." In an age when complex systems give rise to simple patterns and simple systems produce complex behavior,[18] perhaps it is time for diversity to be seen less as a problem and more as an opportunity. There is still ample scope to apply KISS, and this will undoubtedly continue in traditional systems engineering. Given that legacy systems *ab initio* present a given and possibly great diversity, what should the SoS designer do? The purpose of the interoperability framework is to get the legacy systems, holons, to work together, and to do so not additively, as in the current underachieving case, but synergistically. Does this mean reducing diversity, and if so, how can the SoS match the huge diversity in the problematique it faces? The opportunity for the SoS is to increase connectivity, which probably translates into standard protocols and specific architectures or topologies, an imperative for uniformity *and* increased diversity. This respects the autonomy of the holons, allowing them to maximize their contributions to the SoS but within the context of the SoS.

Increasing diversity is not a license for anarchic design, but it is a spur to realizing resilient capability. Situational awareness is enhanced by multiple perspectives. But in the end a common operating picture that informs command decision is just that—a final conclusion. But no one wants to make decisions based on a conclusion that is not richly informed, that is lacking a vital piece of data, information, knowledge, or wisdom. Diversity, through a variety of viewpoints, processes, technologies, and functionalities, ensures richness, and the SoS must be able to leverage this, in an unencumbered fashion.

8.7.5 Emergence

The terms *emergent* and *system* are inseparable. By definition, when parts and their relationships are assembled together what emerges is the system. All systems are emergent. Herbert Simon,[19] a Nobel Prize winner, said this another way when he argued that complex systems will evolve from simple systems much more rapidly if there are stable intermediate forms than if there are not; the resulting complex systems in the former case will be hierarchic.

The properties, behaviors, and purposes attributed to systems can also be said to be emergent. Some of these, for designed systems including the engineered variety, are intended. For example, it is intended that an automobile serves the purpose of transporting goods and people across reasonable distances and terrains safely, comfortably, and in timely fashion. This is an emergent or resulting property of that system. The same emergent property cannot be attributed to any of the parts therein, although every one of these will have its own emergence. Each one is engineered to a specific purpose, to deliver an emergent property; for example, the power train to provide propulsion, the wheels to provide traction, and the steering to provide guidance control. With this example in mind, one can move up and down the scale of systems enumerating specific emergent properties for each part, subsystem, and system.

Some emergent properties are unintended, and of these, some are undesirable and others serendipitous. Relative to the auto, perhaps the chief undesirable and unintended behavior is atmospheric pollution, most acutely experienced in city traffic. At that level traffic jams are another example of unintended emergence: not a single vehicle is responsible for a traffic jam; it takes a bunch of interacting autos to create one. Yet a desirable emergent property at that level is a personal mass transit system, highly convenient if not altogether rapid, one that obviates the need for investment in alternatives such as subways (for cities) and rail networks (for intercity travel).

The question arises, If all systems are emergent, is there anything different or special about a SoS? A SoS must match the agility of the problematique, which calls for greater emphasis on strategic capability than on rigid tactical measures. The exact nature of the SoS is often determined in real time, and indeed at higher clock speed than that of the environment (or the threat within that environment). The simplest way this can be further explained is to draw a comparison between a system and a SoS.

A system provides a response to a set of predetermined requests, that is, threats or opportunities arising from the environment in which it operates. By contrast, a SoS is an anticipatory responder having an a priori undetermined and unknowable range of responses subordinated to auxiliary mechanisms for anticipation, including disturbing the ability of the environment to pose threats or limit opportunity. In the next section we will use a case example of a proclaimed SoS to show how these characteristics may define and realize a SoS.

8.8 Back to Biology

The architecture of DNA comprises a double-stranded helical structure in which the two twisted legs of the ladder are made of phosphates and sugars, and the rungs of nitrogen bases. The work of many scientists contributed to this definition, but it was Watson and Crick[20] who discovered what the rungs were made of and how they joined together and to the legs. From the biological perspective the crucial questions to be resolved were those of growth and reproduction. The architecture settled this, with the particular sequence of nitrogen bases in the rungs containing the coding or instructions for growth and the bonding within bases, limited to specified pairs, conveying the reproductive signature.

What might it mean to think of a systems DNA (sysDNA)? If such existed it could mean that we had access to the vitality of a system in terms of its growth and reproduction. Thus, possibly a means or perhaps the means to adapt it at a genetic level with some guarantee of the outturn, which perhaps we do not normally have by traditional science or engineering. But why should it exist? There are many reasons why it should not. For instance, we know of no body of knowledge for systems that is equivalent to chemistry, physiology, or biology. Perhaps it is mathematics, but the application of this to systems per se, as opposed to models of particular systems, is notoriously lacking. Another primary reason for the nonexistence of a systems DNA is that there is no prima facie reason for it to exist. It is one thing to explore natural systems, since life is what they have in common. But what do designed systems, such as satellites or battleships, have in common that is a mystery we might believe will yield to an equivalent biology, physiology, chemistry, or physics? What are the *Arabidopsis*[21] and Drosophilidae[22] in the study of systems? How can so very many ad hoc man-made designs possibly bear a common imprint that would suggest an underlying DNA. It is unlikely if not inconceivable.

Then for systems, which are mostly technology intensive, can we define the fundamental building blocks by which we can then apply a systems study to? One thing that systems do have in common, or at least it is said they do, is an architecture; however, this is codified and there are multiple instantiations. What this does convey is conceptual design from which the detailed system later takes its form and thence its function. Is system architecture the systems DNA? Unlikely, since it is almost certainly specific to each system, although patterns might be observable across the spectra of system architectures, certainly across given types of system. So somehow or other the system architecture may have something to do with a systems DNA, but revealing the independent existence and nature of the latter is still what we desire. If it exists. Traditionally, the attempts to understand the life cycle of systems have been through the engineering of systems or systems engineering. But systems engineering provides us with a process by which we can design, develop, and manage systems, and does not tell us what makes up a system. What are the systems building blocks, so we can define and study the biology of all systems?

We want to be able to rely less on reductionism and discovery to understand systems, and move to a more hypothesis-driven and discovery science approach to systems. Most systems are fundamentally complex with multiple parts, and it is almost impossible to fully understand and study something with so many parts. In reductionism we would reduce that down to the lowest level possible, but leaving ourselves with the improbable task of trying to relate this to the original problem. Systems biology has attempted to apply general systems theory to being able to explain the integrated and interactive nature of biology from the molecular level to the macro level. The study of systems has no molecular level; there is no systems cytology. Therefore, we are asking the fundamental but not trivial question, What is the DNA of systems?

The cell and a system have three things in common: structure, function, and life cycle. The cell has the advantage of chemistry to explain structure. The system does not. The cell has its vitality encoded chemically. But this encoding essentially deals with patterns of organizational form to which we can usefully impute anthropomorphisms. Carbon, for example, is very agreeable, more so than any other element. It happily builds relationships with other elements. Some architectural forms are rugged, difficult to break. Others are fragile. And yet both types may consist of the same elements, just differently arranged. This line of thinking gives us a clue as to how to build a conceptual chemistry and with that the fundamental building blocks of systems such as satellites, battleships, and the product development teams that build them. We believe that the essence of a system is togetherness, the drawing together of various parts and the relationships they form in order to produce a new whole that will have its own structure, function, and life cycle, as we discussed in the previous chapter. That said, as described in Section 8.5 and shown in Figure 8.4, we want to embark on a discovery of togetherness, a conceptual chemistry that consists of the five characteristics as elements.

For each of these elements, there are opposing forces or paradoxes that are influenced by fluxes in realizing or recognizing a system (we will discuss the nature of paradox further in the next chapter). This balance is considered reversible. Reversible is conditions under which the forces are so nearly balanced that an infinitesimal change in one or the other would reverse the realization of the system. In any system we seek ideal conditions that the realization of the system is carried out reversibly. Under these conditions the realization of the system yields the maximum possible performance, although reversibility does not hold true in practice. The flow of these forces and their relationships works in distinguishing types of systems and determines the togetherness of a system, which fortifies its realization. In biology, energy plays a fundamental role in the chemical and physical processes that help to realize all living systems. Energy and the principles of thermodynamics have long provided a deeper understanding of the interrelationships between structure;[23] therefore, for systems, we contend that this energy is to biology as togetherness is to systems.

Today biologists from myriad domains of specialism are once again seeing the virtue of systems thinking and its fundamentals as evidenced by the wealth of interest in systems biology.[24] We are attempting to provide greater formalism to the notion of system ubiquity, that is, to describe a system in the abstract so that system designers and managers of specific systems can take account of this abstract knowledge, thereby ensuring that whatever they build is a system not merely because it carries that term in its description, but also because it bears the marks of a system as we understand that term. We rely on the notion that a system is a collection of parts and their interrelationships assembled together to fulfill a purpose. Our differentiating elements have something to say about these parts, their interrelationships, the assembling together (process), and the fulfillment of purpose—all in the most abstract sense, but in a way that this relates to the specifics of the system under consideration.

8.9 Time to Think

1. There once were two watchmakers, Hora and Tempus, who manufactured very fine watches. Both of them were highly regarded, and the phones in their workshops rang frequently—new customers were constantly calling them. However, Hora prospered while Tempus became poorer and poorer and finally lost his shop. What was the reason?

    The watches the men made consisted of about a thousand parts each. Tempus had so constructed his that if he had one partly assembled and had to put it down—say, to answer the phone—it immediately fell to pieces and had to be reassembled from the elements. The better the customers liked his watches, the more they phoned him and the more difficult it became for him to find enough uninterrupted time to finish a watch.

    The watches that Hora made were no less complex than those of Tempus. But he had designed them so that he could put together subassemblies of about ten elements each. Ten of these subassemblies, again, could be put together into a larger subassembly, and a system of ten of the latter subassemblies constituted the whole watch. Hence, when

Hora had to put down a partly assembled watch to answer the phone, he lost only a small part of his work, and he assembled his watches in only a fraction of the man-hours it took Tempus.[25]

This story exemplifies the virtue of hierarchical design. Are there any disadvantages to this paradigm? Do some of these disadvantages lie beyond evident advantages, such as the ability to outsource certain modules of a product breakdown structure? Can you envisage a hybrid approach, consisting of hierarchies and networks that provide superior advantage? Discuss.

2. The following is an exercise in network evolution. Consider the *movie A Few Good Men*. Choose five stars from that movie. Treat these stars as hubs of an emerging network. Add one (primary) actor to each star who did not play in the *Good Men* movie. Connect these new actors to any of the hubs on the basis of their playing together in the same movie. Add one (secondary actor) to each primary actor via movies that they played in together. Connect the secondary actors to other primary actors and to hubs on the basis that this likely pairing played in the same movie. Continue one more time with (tertiary) actors and their links with existing nodes in the network. How many links do you have? How many movies did you use? What is the topology of this network? Are the original hubs actually hubs in this network, or are they peripheral? If we used this evolutionary method using corporations and not actors, and contracts rather than movies to be the network links or ties, what lessons could we learn about the corporate networking world and the well-being of a corporation on any given landscape? Given that corporations are essentially hierarchical in nature yet the landscape is not, are there any implications for corporate governance?

3. Let us assume that every individual adult has a set of competences and some communication skills. The competences represent what that individual can accomplish on her own without recourse to any other person. The communication skills represent an ability to relate to others, accept information and instruction, and issue reports and directions to others. Consider a group of individuals, say, five, engaged in some specific purposeful activity, for example, setting up a wilderness camp in readiness for a party of twenty young teenagers. What competences need to be present? What kinds of communication exchanges do you envisage? What conditions are likely to promote the enhancement of competence and the improvement in communication skills? What kind of balance between competence and communication is most likely to be successful, or does it all depend? What are the kinds of errors and misconception that you foresee emerging among individuals that may seriously inhibit the group's achievements? Are there any lessons to be gained from this exercise that could be successfully transferred into corporate life?

4. Returning to the exercise above, use what you learned to refine the five characteristics proposed in the chapter, that is, autonomy, belonging, connectivity, diversity, and emergence, to differentiate a system of parts from a system of systems. For each characteristic, think of three or four words that support the meanings of each and whose togetherness translates into the meaning of the characteristic itself. For example, in the case of autonomy these supporting terms might be core competence, self-starting, self-control, and task ownership. Create a concept network linking these supporting terms, in which the linkages define influences, in both directions possibly, with relative weightings on the degree of influence. Find ways of exercising this concept network using whatever scenarios and exemplifications you conjured in the wilderness camp preparation task. Does this kind of modeling and simulation begin to indicate ways of understanding the togetherness of a system, independent of the function of that systems, that is, the specifics of purposeful activity?

5. Use the distinguishing characteristics presented in this chapter, that is, autonomy, belonging, connectivity, diversity, and emergence, to make judgments as to which of the following is a system (of parts) or a system of systems: concert orchestra, jazz band, ant colony, United States of America, a Boeing 787.

6. In the case of most capital goods, for example, communications satellite, battleship, gas turbine engine, and ballistic missile, the majority of the value lies outside the corporation that manufactures them. In the case of a Trent engine that Rolls-Royce designs and builds, this value lying in the supply base can exceed 70%. It is this kind of consideration that supports the notion of an extended enterprise, that is, a collection of autonomous corporations that choose to belong, to form their respective connections and exercise their diverse core competences to emerge a meta-corporation in order to realize the final product. Is an extended enterprise a system of systems, in which the constituent systems are the member corporations, and not simply a system? If this is the case, how can we make sense of governance and resilience? What might it mean for the many to be in charge as opposed to "the man?" How can the resilience of the constituent systems be maintained when this might lead to a lack of resilience of the extended enterprise? If a significantly minor system, that is, low down in the value chain, possesses a critical technology, how can the extended enterprise safeguard against its vulnerability to adverse cash flow or loss of critical mass to protect its vital contribution, without taking away its autonomy? What other issues come to mind simply by viewing this extended enterprise as a new kind of system?

7. In their book *The Starfish and the Spider*,[26] Ori Brafman and Rod Beckstrom compare and contrast two distinct organizational forms. They point out that if you cut off a limb of a starfish, two things happen: first,

the starfish grows another limb, and second, the limb itself can become a new starfish. Interestingly, they liken Al Qaeda to a starfish. If we were to mistake Al Qaeda for a spider and attempt to cut off its head, notionally Osama bin Laden, which is meant to have the effect of killing the spider, what actually will occur is growth of the enemy. This is a typical counterintuitive result that regularly occurs when you have a misguided systems perspective. Here are some questions to consider: What would determine Al Qaeda to be a starfish? How is the DoD organized? What should our approach be to neutralizing Al Qaeda? What do we need to do, to ourselves, to be an effective opponent? Is there any similarity between these organizational forms and the two *of*'s presented in the chapter—hierarchy and network? What would a hybrid form of *of* be, and equivalently, an organizational form that was a hybrid of starfish and spider? And how can a social organization, especially one that has a heritage in military service, become a hybrid enterprise?

8. What is emergence? What is an emergent property? Do all systems have emergent properties? Are all emergent properties surprises, or can some be planned? Are all surprises beneficial, or are some debilitating? If emergent properties can be planned, what form of planning safeguards against unwanted or undesirable emergence? If emergent properties are both unforeseen and unforeseeable, what planning strategy is possible, if any, to cope with emergence? Finally, what is the essential difference between emergence in a system (of parts) and an emergence culture in a system of systems? How and under what circumstances would you want a system of systems for its kind of emergence?

Endnotes

1. Surowiecki, J., *The Wisdom of Crowds*, Anchor, New York, 2005.
2. DiMario, M., "System of Systems Characteristics in Joint Command and Control Interoperability," in *IEEE International System of Systems Engineering Conference*, Los Angeles, April 24–26, 2006.
3. Doz, Y. L., and G. Hamel, *Alliance Advantage: The Art of Creating Value through Partnering*, Harvard Business School, Boston, 1998.
4. Sage, A., and C. D. Cuppan, "On the Systems Engineering and Management of Systems of Systems and Federations of Systems," *Inf. Know. Sys. Manag.*, 2, 325–45, 2001.
5. Carlock, P. G., and R. E. Fenton, "System of Systems (SoS) Enterprise Systems Engineering for Information-Intensive Organizations," *J. Sys. Eng.*, 4, 242, 2001.
6. DeLaurentis, D., "Understanding Transportation as a System of Systems Design Problem," in *43rd AIAA Aerospace Science Meeting*, Reno, NV, January 10–13, 2005.
7. Keating, C., et al., "System of Systems Engineering," *Eng. Manag. J.*, 15, 36–45, 2003.
8. Luskasik, S. J., "Systems, Systems of Systems, and the Education of Engineers," *Artificial Intel. Eng. Design Anal. Manuf.*, 12, 55–60, 1998.

9. Lane, J. A., and R. Valerdi, "Synthesizing SoS Concepts for Use in Cost Estimation," in *IEEE International Conference on Systems Man, and Cybernetics*, October 10–12, 2005, 993–98.
10. Bar-Yam, Y., "The Characteristics and Emerging Behavior of System of Systems," *NECSI: Complex Phys. Biol. Soc. Sys.*, 1–16, 2004.
11. Boardman, J., and B. Sauser, "System of Systems: The Meaning of Of," in *IEEE International Conference on Systems of Systems Engineering*, Los Angeles, April 24–26, 2006, 118–23.
12. Sauser, B., and J. Boardman. "From Prescience to Emergence: Taking Hold of System of Systems Management," *27th American Society for Engineering Management National Conference*, October 26–28, 2006, Huntsville, AL.
13. Ibid.
14. Smuts, J. P., *Holism and Evolution*, Macmillan, London, 1926.
15. Koestler defines holarchy to be a hierarchy of self-regulating holons that function (a) as autonomous wholes in supraordination to their parts, (b) as dependent parts in subordination to controls on higher levels, and (c) in coordination with their local environment. See Koestler, A., *The Ghost in the Machine*, Penguin, London, 1990.
16. Stenbit, J. P., L. Wells, and D. S. Alberts, *Complexity Theory and Network Centric Warfare*, DoD Command and Control Research Program, Washington, DC, 2003.
17. Meaning perhaps both the similarity of retaliation and the willpower/passion behind it.
18. Waldrop, M., *Complexity: The Emerging Science at the Edge of Order and Chaos*, Simon & Schuster, New York, 1992.
19. Simon, H. A., *The Science of the Artificial*, MIT Press, Cambridge, MA, 1996.
20. Watson, J. D., and F. H. C. Crick, "Molecular Structure of Nucleic Acids: A Structure for Deoxyribose Nucleic Acid," *Nature*, 17, 4356, 1953.
21. *Arabidopsis* is a genus in the family Brassicaceae. This genus is of great interest since it contains Thale cress (*Arabidopsis thaliana*), one of the model organisms used for studying plant biology and the first plant to have its entire genome sequenced. Changes in the plant are easily observed, making it a very useful model.
22. Drosophilidae is a family of flies, including the genus *Drosophila*, which includes fruit flies, vinegar flies, wine flies, pomace flies, grape flies, and picked fruit flies. These flies are used extensively for studies concerning genetics, development, physiology, ecology, and behavior because of very short life span and gradual aging.
23. Francios, C., "Systemics and Cybernetics in a Historical Perspective," *Sys. Res. Behav. Sci.*, 16, 203–19, 1999.
24. Kitano, H., Ed., *Foundations of Systems Biology*, MIT Press, Boston, 2001.
25. See http://easyweb.easynet.co.uk/~iany/consultancy/historical/historical.htm for further analysis.
26. Brafman, O., and R. A. Beckstrom, *The Starfish and the Spider: The Unstoppable Power of Leaderless Organizations*, Penguin, New York, 2006.

chapter nine

Paradox

9.1 Make My Joy Complete

Once upon a time there lived a farmer and his three sons. Together they were a family. The farmer loved his sons and they adored their dad. He taught them all they knew. The family had a neighbor who loved the sons like they were his very own, and the farmer knew that he could count on his neighbor should anything happen to him. And indeed it did. The farmer died, leaving his sons as orphans. But not alone.

In his will the farmer left his entire herd to the three boys, seventeen cows in all. He did not, however, distribute this total equally, for which the boys were grateful. To the eldest the farmer bequeathed half the herd. To the middle son went one-third of the cows. And to the youngest boy exactly one-ninth was given. This distribution left the boys frustrated and confused. Much ran through their minds, as you might imagine, and their consternation did not escape the neighbor's attention, who dutifully and lovingly continued to watch out for their well-being just as the old farmer had hoped.

"What is troubling you?" asked the neighbor. They explained the terms of the will and their dilemma in dividing up the herd bloodlessly. Thinking that their only recourse was to sell the herd and apportion the proceeds as best they could, the neighbor suggested an alternative. He went away leaving them with the intriguing remark "I'll be back!" to puzzle on during his brief absence. (In Chapter 2 we offered this as an exercise. Now we give you our answer, only to confront you with yet another deeper puzzle!)

The neighbor returned with a cow. The only one he had. A straggly looking beast, hardly suitable for anything and incomparable to the quality cows their father had raised. "I think this may resolve your dilemma," said the neighbor with a faint smile. At this point the boys thought their neighbor had taken leave of his senses. It was a challenge to their humility to accept such a pathetic gift. But their dad had brought them up well. So they accepted the offer, trusting that the neighbor, whom the father had loved as his dearest friend, would not fail. Now the boys had eighteen cows, considering that this was not only their father's will but also in his will. The eldest boy took his nine cows. The next in turn took his six and the youngest his two. All the boys were careful not to include the recent addition in each of their picks. Strangely, the straggly bovine remained unclaimed.

The neighbor asks, "Are you all now satisfied?" They nodded in vigorous unity. The terms of their father's will had been observed with complete fidelity with not one drop of bloodshed—of neither beef nor brother—as the old farmer had always intended. "I guess I'll go on home. It's late," said the neighbor, who slowly headed off with the discarded scrawny animal in tow and intact, its purpose served.

Our question to you is: How can this unwanted animal be needed *and* unneeded—*both* at the same time? Further, what is the meaning of simultaneously applicable opposites? And what have such questions got to do with systems thinking, especially its application to technology development and to enterprises that either execute such development programs or are themselves enabled by the products of technology?

9.2   The World of Both

A dictionary definition of paradox is "a statement that contradicts itself." For example, consider the statement "Even one of their own prophets has said, 'Cretans are always liars, evil brutes, lazy gluttons.' This testimony is true."[1] This represents a paradox. If a Cretan says "All Cretans are liars!" how can this be true? If what the Cretan says is true, it must mean that he is a liar and therefore his remark cannot be true. Yet if indeed all Cretans are liars, then the Cretan cannot truthfully say what he says, making the premise "all Cretans are liars" false. Round and round we go, world without end.

Consider a second example. On a single sheet of paper appear two statements, one written on each side. On the first side we find: "The statement on the other side of this paper is true." On the other side we have: "The statement on the other side of this paper is false." More endless circularity. Does this lead us anywhere?

Here is what we have to say about paradox:

> A paradox is an apparent contradiction; however, things are not always as they seem. A paradox can be explained, but only by seeking wisdom from above; for the systems person this means looking upwards and outwards, not just down and in. Paradoxical thinking is systems thinking at its best.

When we use the word *apparent* we give ourselves the opportunity to introduce viewpoint or perspective, which is personal or subjective to some individual (person or group), and this notion, sometimes labeled stakeholder, is germane to systems thinking. Further, we hold out the hope of resolution as opposed to the sense of despair or confusion that can befall those who get trapped in the endless circularity of evident paradoxes. Finally, we point to a source, as Curly did (and didn't) in *City Slickers*.[2] We locate that source as "from above" and denote it as "wisdom." It is for you to determine what *above* and *wisdom* mean, as we must also do and have done for ourselves.

Many apparent paradoxes that once confused the greatest minds of the day were later explained by new concepts and new ways of thinking. One of Zeno's paradoxes[3] that confounded the ancient Greeks was not resolved until Newton and Leibniz conceived of the limit, and with that invented calculus, and discovered irrational numbers. That paradox was seen as the conflict of the irresistible inference and the inescapable fact, for some an elegant contemporary definition of paradox.

The essence of paradox is *tension*—two statements claiming to be true and at the same time contradicting each other. The ultimate release of that tension, not found in the resolution of the conflict within the paradox itself but rather in the recognition of the virtue of the paradox as a whole, always leads to new ways of thinking. For this reason, as much as paradox is unpalatable, especially to action-oriented people such as engineers, technologists, and business executives, it can be valued as a lever to change mindsets, to shift thinking, and a potential wellspring of new ideas leading to more effective action.

In this chapter we will explore a variety of paradoxes and types of paradox taking care at all times to gain an understanding of the underlying tensions. We assert that to resolve this tension prematurely, for the sake of taking action (lest it be too late!), always leads to nugatory action and, more expensively, to missing an opportunity to change your way of thinking and gain breakthrough knowledge. Likewise, to ignore this tension, to pretend as it were that the paradox is unreal or irrelevant, is to preserve the status quo, maintain the same old grid lines of thought, and inevitably head for disaster.

Paradox is a reality of our lives. Tension is too. We have chosen to feature paradox as a significant element of systems thinking because we see paradox less as a source of confusion, which at face value it certainly is, and more as a portal into new ways of thinking, new modes of working, and better ways of living.

We want to explore what we call the world of both. In a paradox there are two opposites that compete for our attention, and each demands we make it as our choice.[4] But the very existence of the paradox itself demands that we do not choose, but that we accept both, and therefore the nature of the relationship between the parts, manifested as a conflict, contradiction, or something other than runs counter to common sense or conventional wisdom. Systemically, the parts of the paradox demand choice, but the whole of the paradox requires acceptance of both, as illogical as that seems. It is as if parts and whole cannot agree, and yet parts make the whole and parts they be.

In our experience neither engineers nor managers are comfortable with the notion of both, of holding on to a tension that *must* be resolved. They like choice, perhaps not an overabundance of choice, and they are required to choose. It is their raison d'être, their nature of being.

We believe that systems engineers and systems managers need to be, or learn to be, comfortable with *both*. Scientists have had to learn to accept subatomic matter to be *both* particle and wave, while technologists in pursuit of the quantum computing dream postulate the qubit,[5] a binary digit that is *both* one and zero, and consequently are able to leverage computing power exponentially. Both, if not as yet in, is on the scene, and we are offering our own thoughts and examples to make it easier for decision makers to be cool with simultaneous opposites (aka conflicting perspectives) leveraging these into richer realms of decision making.

9.3 System Paradoxes

In this section we want to offer you some thoughts on the occurrence of paradox relative to a selection of the key systems concepts that we touched upon in Chapter 2. We have found it surprisingly easy, albeit alarmingly uncomfortable, to discover these various paradoxes. We had always thought of systems thinking as a source of strength and a capability to live with dilemmas, ultimately resolving them. Au contraire! We find this body of knowledge to pose more questions and present more challenges than it purports to address. Then again, maybe that is its true purpose.

9.3.1 Boundary

The notion of boundary is inseparable from that of system. As problematic as it might be to locate or articulate the boundary of a system, that it exists relative to the system itself is incontestable. The boundary may be defined by geography or other dimensions such as culture, organizational structure, or IT infrastructure, but howsoever it is defined, it fundamentally speaks of separation, of distinction, of limitation, as well as approximation. The boundary shows who and what is in and out.

If a system be a collection of parts and their interrelationships assembled together to form a whole for a given purpose, then the parts and their relationships are in (the system). It is they that are together, or intendedly so. What lies beyond the boundary is not part of the system. It may be coveted by the system and in due time be acquired and integrated into the new whole, but while it lies outside the boundary it cannot be considered part of the system and may even be considered hostile to the system.

The parts of the system can be controlled so as to serve the system or, in self-organizing style, relied upon to control themselves autonomously and bring even greater well-being to the system. The externalities cannot be controlled; they may even need to be combated, if regarded as foe rather than friend. Possibly they can be influenced, thereby rendering docile an otherwise adversarial influence on the system itself.

System designers must be perpetually mindful of the boundary. If it is an investigation into an airplane crash, they need to decide where to look, intently and otherwise, and what can be safely discarded. If it is a piece of electrical equipment, the designers may need to pay special attention to electromagnetic compatibility (EMC), which will have consequences for the system boundary relative to both emission and immunity issues. If it is a nation-state, the system boundary has influence on regulating trade, patrolling for unwanted "visitors," and currency exchange.

Chapter nine: Paradox

So the system boundary separates what belongs to the system from what does not belong. And in the system's best interest, only the good must belong and the bad must not; otherwise, it gets ugly. The more good available, the better for the system; the less bad, the better. In the movie *A Few Good Men*, Daniel Kaffee[6] (played by Tom Cruise) asks his buddy Sam Weinberg (Kevin Pollack), "What's a fence line?" to which Weinberg replies, "It's a big wall that separates the good guys from the bad guys." That nails it. Or does it?

As much as the human body prefers not to be invaded by unwanted bacteria, immunization (letting a specified amount of bad guys in) enables the good guys to get better at dealing with the bad guys so when large numbers of these try to invade, the body's defenses have significantly improved. Does this mean a certain number of Mexican illegal immigrants is a good thing? And if so, how many should that be? And what does that mean in terms of secure border?

And is it such a bad thing to lose good guys from a system? Think about Lockheed and its famous Skunk Works®.[7] In a very real sense the Lockheed Corporation deliberately cut loose its A team, freed them from the increasing bureaucracy and supported them in getting critical products out the door. These folk were a "loss" to the company, but Skunk Works was a gain to Lockheed. Conversely, when a few guys from Fairchild Semiconductors left the company, because it would not respect their suggestions for the fabrication of new integrated circuits, they formed a tiny start-up company by the name of Intel. Those guys were a loss. Fairchild no longer exists, and the fortunes of Intel's founders can be counted in the billions.

Our boundary paradox can be stated as follows: You have to have a boundary (in order to nurture and develop specialization in functional expertise, for example). *But* you must also *not* have a boundary (in order to allow that specialization to be rendered as a service, otherwise why have it, and to allow that expertise to be resourced via interactions with others). So the boundary must exist and must not exist—it must do *both*, at the same time. The boundary must keep things out and keep things in, but it must also let things out and let things in. As far as the human body is concerned, this makes the operation of the cell wall (see Figure 9.1) a minor miracle.[8]

Society owes a great debt to electrical engineers who have made this existent/nonexistent possible when it comes to electrical signaling. The same copper wire, a kind of bridge that spans the boundary between two communicants, can simultaneously support signals in both directions—something termed full duplex.[9] In that sense the paradox of the boundary has been resolved, but on closer inspection it has not. Whatever multiplexing techniques are used to handle simultaneity, they rely on dividing up time, frequency, or bandwidth to accommodate the signaling. The metaphor of two-lane highways is apposite, but what have been created are two boundaries—one to admit and the other to exclude; after all, vehicles traveling in opposite directions do not use the same highway lane simultaneously. And, in fact, all of our resolutions of this kind of paradox boil down to serializing

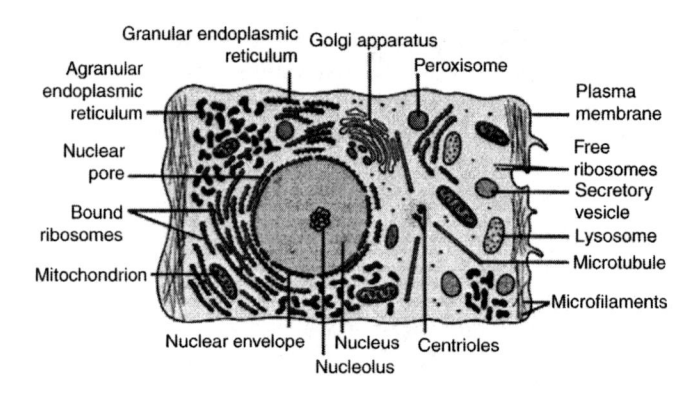

Figure 9.1 Human cell. (From Vander, Sherman, Luciano, Human Physiology, 7th ed., 2007. Reprinted with permission of McGraw-Hill.)

an essential parallelism. This is okay practically, but it can be very unsatisfactory philosophically since it obviates the need to recognize the boundary paradox for what it is: a fundamental tension between opposing forces operating simultaneously; inside and outside forces and forces for good and evil. The release of that tension, the ultimate resolution of this paradox, leads to thinking of higher orders. A realm we ourselves have yet to enter but to which we journey.

9.3.2 Control

How many people and how many firms do you think are involved in the end-to-end process of conceiving, making, and selling a Grand Cherokee Jeep? From initial product concept through to a satisfied customer driving her new purchase off a dealer's lot. It is perhaps not a question that interests many, but the answer usually startles most.

When Thomas Stallkamp, former VP of Chrysler, asked this question of his line managers, it took them a little while to find the numbers. In the end they came back with the answer: 100,000 firms and 2 million people. Wow! But once you have gotten over the shock, here is Stallkamp's follow-on question: "Who's managing this enterprise?" The real answer is a paradox: no one is *and* lots of people are. Yet another time to make your mind up? Or a time to recognize and respect the paradox, waiting until the appropriate moment to release the tension and to achieve breakthrough thinking.

No one sits atop the Chrysler Grand Cherokee Jeep "experience." Perhaps notionally someone does, but in no way can he or she be said to be its manager. Whoever occupies that office is no George Washington or Emperor Napoleon commanding thousands and controlling affairs according to his or her grand strategy. Littered throughout the management hierarchy, or network if you prefer, are hundreds of personnel each with their individual

Chapter nine: Paradox

spans of care. But in what ways can this diverse collective be said to be in control of the whole experience when it is probably the case that they are largely unknown to one another? Do these managers perform like ants and somehow support excellent behavior for the Cherokee colony? And if so, understanding that the ant has no commander directing colonial affairs,[10] are we to understand control to be just as effective, if not more so, if it is distributed rather than precisely located in a central commander? And can we really trust distributing control to a constituency that is largely unaware of the affairs and actions of its neighbors?

In the movie *Bobby*, Lawrence Fishburne has a great line: "The white man finally gave us our freedom. But it was never theirs to give. We just let them think it was." So when we ask "Can we really trust distributing control to …," maybe we are asking the wrong question. Just like when Stallkamp asked who (in particular rather than plurality) is in control of this vast extended enterprise; maybe he was asking the wrong question. We ask the wrong questions when we are in the wrong mindset. And the purpose of paradox is to confront that mindset. To force us to ask wrong questions. To stop and think: Maybe we are asking the wrong questions. And to be prepared to change our mindsets, thereby releasing the tension in the paradox and moving to breakthrough thinking.

So what are the right questions? Well, stepping back: What is the right mindset? Maybe control is, or at least starts with self—self-control. After all, you have to exercise self-control in response to an order, be it in the military, civil, or family domains. What is more, the one issuing the order expects this, relies on this self-control, and in some way is developing this in the one to whom the imperative is directed. Command assumes self-control. The question now is, Is there an extension to command that is of a controlling nature, that is, the communication of command carries with it or in it a controlling influence? This is at the heart of resolving strategy into tactics. Churchill's mantra (his strategic goal) was "Liberate Europe." But to issue this as a command to millions is a little pointless, especially in the face of an organized, commanded, and highly controlled Nazi Party and its associated military might. But to issue it as an idea is not a bad idea. Under the right conditions and at just the right time an idea can infect people, in increasingly large quantities, and this can lead to, for example, huge market dynamics (cell phones) or even political revolution.

At this point we are beginning to recognize the polarities. One is the command and control version by which authority located "at the top" issues directives that get resolved into executive action by a large group of people. The other is the self-organizing notion of an idea that (from the bottom) infects, propagates, and galvanizes a large group of people who then take action, as though they were a unit and had been commanded by a governing authority. Some are calling this infecting idea a meme, impersonating the gene notion.[11]

It is this existence of polarities that leads us to formulate our second systems paradox: You have to have command and control (in order to ensure orderliness and conformity to strategic direction). *But* you must also *not* have command and control (and instead have ground-zero intelligence to foster innovation, tactical opportunism, and preservation of self-awareness).

Put in other ways:

- Authority must exist at the top representing order, but it must also exist at the bottom representing autonomy.
- Command must exist and orders from an external source be obeyed, but so also must the power to be insubordinate operating alongside a self-will that knows its own order and orders.
- Finally, control must operate within a framework (a one) that grants liberty to its constituents (the many), but control must also be manifest in the self (a one) in terms of self-control and self-discipline to make a framework (for the many) work.

What are the ways in which this paradox might be resolved and its tension released? We touch, briefly for now and more fully at the close of this chapter, upon three measures: creative disobedience, reciprocal loyalty, and ordered liberty. These are notions associated with no less a commander than Admiral Horatio Lord Nelson, and the notes reproduced below are attributed to Leigh Kimmel.[12]

> The idea that an individual commander as the man on the spot should have the flexibility to deal with the situations as they came was a central part of Nelson's battle doctrine. He had a talent for communicating his ideas and plans to his captains so well that they understood what he would want them to do in any specific battle situation and carried it out as well as though he were there. Thus he was able to keep his orders general. In his orders for the assault in Tenerife, item six noted that his captains were "at liberty" to send more men and to land under Troubridge's direction rather than have to get specific orders from Nelson. He also had the battle plan for the Nile worked out almost two months before he actually entered Aboukir Bay. His was the master plan and he left the details to individual captains, believing that they had the good common sense to innovate and act independently. Foley's decision to go inside the French line at the Nile when he saw the opportunity fits perfectly with Nelson's philosophy of independence of command. Nelson's orders to his captains at Copenhagen were also quite bare and simple. He expected them to apply these general details to the specific situations they encountered as the battle unfolded. Finally, Nelson's famous memorandum

circulated before Trafalgar gave Collingwood full latitude to fight his whole line as necessary.

The concept of **creative disobedience** flowed naturally from his philosophy of independence of command. If his subordinates should have the freedom to deal with situations as they came up, he should be able to take the initiative as a subordinate in a battle, even if it meant ignoring orders. The first great example of this was his "famous indiscipline" at the battle of Cape St. Vincent, where he pulled out of the line of battle in order to interdict the Spanish flagship, thus allowing the rest of the British fleet to catch up and get into fighting position. However this involved breaking the standing orders that no ship was to leave the line of battle without permission from the senior admiral. Oliver Warner claims that no other subordinate officer has taken such an initiative as Nelson did at Cape St. Vincent, although he does not make clear whether he is comparing Nelson only to other officers of the Royal Navy or officers of all navies (which would also require examining the history of the Pacific Fleet in World War II, which had its fair share of gung-ho admirals who didn't always mind Nimitz). Sir Nicholas Harris Nicolas, editor of Nelson's letters, suggested that Jervis didn't praise Nelson for his success because Calder, Jervis' flag captain, pointed out that Nelson had disobeyed standing orders to stay in the line of battle and that praising such disobedience would set a bad example for future officers. However Jervis is recorded as having responded to Calder's criticism with the remark, "... if ever you commit such a breach ... I will forgive you also."

Reciprocal loyalty is the idea that one must give loyalty down the command hierarchy in order to gain true loyalty (as opposed to obedience through fear). Nelson seems to have understood this instinctively, although his year of service on a merchant ship at the very beginning of his career may well have helped to shape that instinctive understanding into practical action. His career shows many examples of the way in which he stood by his subordinates and saw to their welfare. Near the end of his years ashore, between the time of the storming of the Bastille and the execution of Louis XVI, certain elements of English society were becoming restive with the possibility of freedom promised by the French Revolution. While many aristocrats and country gentry were responding with hysteria, Nelson started looking for the cause of the problem and its solution. He went around the Norfolk countryside talking to ordinary people about their grievances and put the knowledge he gained to work. He also did his best to improve conditions aboard his ships and to see to the welfare of the sailors under him and their families. After the Battle of the Nile, Nelson wrote a letter to Lord Spencer (then First Lord of the Admiralty) asking after the welfare of the fourteen-year-old eldest son of a Marine officer who was killed aboard his flagship in that engagement. After the Battle of Copenhagen he wrote several letters trying to get recognition for his brave followers. In a letter to St. Vincent he expressed his belief that the commanders at Copenhagen should be given medals. In a letter to the Lord Mayor of London he claimed that he wouldn't complain if his reputation were the only thing involved, but he had the bravery of his subordinates to consider and wanted them recognized. And shortly before the Battle of Trafalgar the bosun who loaded Victory's mailbags forgot to include his own letter home to his wife. When word of this got to Nelson the mail ship was already a good way out, but the admiral called it back to pass the one letter, remarking that the bosun might well fall in battle the next day. These small concrete actions won his sailors' love in a way that no amount of grand speeches and posturing could ever have.

Our two questions to you are: First, do these measures address the paradox of control, and if so, in what sense do they constitute a breakthrough in mindset? Second, maintaining the biblical references we have been making in the chapter thus far, and only if you are of a mind to do so, to what extent do these notions support that of *servant leadership* so supremely exemplified in the inspirer of the Christian faith whose teaching is incomparably paradoxical?

9.3.3 Diversity

Together each achieves more. It is a neat phrase. It epitomizes togetherness. It gives a sense of fulfillment that is somehow eluded by the mindset of going it alone. It seems to make selfishness redundant and self-achievement more rewarding because self is being helped by others and self is helping others simultaneously. It also conjures the notion of being coached or mentored or somehow developed, as a consequence of which life is more rewarding, learning is gained, and transferable skills acquired. No wonder people use this phrase as an acronym for TEAM. It is almost so engrained nowadays that you cannot be in a team without realizing that while more is the goal of each and everyone, it comes at the expense of being together.

A team simply has to be a system. It may be a poorly performing team and therefore a failing system, but a system nonetheless. It is worth our while spending some time looking at what a team is and what it means, as an example of a system, to discover yet another interesting and rather fundamental paradox that can so easily go unnoticed by system designers and operators, as a consequence of which we have more bad systems than we need.

In a team we find both sameness and differentiation. Sameness is exhibited in uniformity: a baseball team has its colors, its motif, its insignia, its war cry, and its nickname. Everybody on the team and associated with the team identifies with these unifying themes and artifacts. They become recognizable and identifiable. They help to define the personality of the team distinguishing it from other teams. It is an emergent oneness that covers the many identically. Sameness is also exhibited in the common aim of success, of achieving more. Each is signed up to this notion, and anybody found wanting is quickly identified and almost certainly removed. A team has no room for mavericks, loners, rebels, dissidents, and the like, regardless of individual expertise, no matter how exceptional. A member not committed to togetherness cannot be on the team. His or her untogetherness is an automatic disqualification. The former sameness, corporate identity, is physical and tangible. It is evidential. It has an objective reality. The latter sameness, team success, is more intangible. Indeed, it can be quite subjective; different team members have differing views on what constitutes team success. Ironically, this apparent sameness can be highly distinctive, though improbably divergent. But this distinctiveness is unimportant because at a deep-rooted level the sameness, the single commitment to team success and achieving more together, is overpoweringly unifying.

From time to time the sharp distinctions in subjective interpretations of togetherness can be a powerful disintegrating force. Ultimately the team is debilitated and drastic action might be required by significant stakeholders to get the ox out of the ditch. That is the price of differentiation operating simultaneously with sameness. The tension between the two is, in this instance, disruptive in the most unhealthy fashion. However, patterns exist to provide early warning signs, detect the potential demise, and make timely interventions.

But differentiation is essential for many reasons. Sameness is needful. But not singularly so. A team is a blend of many skills. In baseball you need pitchers, basemen, outfielders, batters, and runners. In soccer you need goalkeepers, goal scorers, goal scoring providers, and defenders. In an engineering team you need people from different disciplines—electrical, electronic, mechanical, and software. You also need people with different project experiences—in leadership, in work package management, in test, and in manufacturing. Teams need different skills, knowledge bases, and experiences. The team becomes a pool for blending these differences together, so that each achieves more.

A team simply has to be a system because it has to have requisite variety, that is, differentiation, parsimony (a meanness that culminates in the single-mindedness of each member to put the team first), and harmony (that which makes togetherness feasible).

What is true for a team of individuals is true also for a collection of technological components gathered together to form a new system—there is both differentiation of functionality and of sameness, performing the functions on behalf of achieving system purpose, of fulfilling the system's mission. But before we turn to traditional systems, let us stay a while with the notion of team, what is increasingly being labeled the enterprise system. Let us ask the following questions of an individual team member: What does it mean to you to be *both* different and same at the same time? Does this present a dilemma? Are you faced with a paradox? At first glance the idea that these simultaneous opposites create a paradox would not arise. The individual is a member by virtue of doing his or her job and the team benefits from this individual doing his or her job and in so doing by being a member. It is as simple as that. End of story. Or is it?

There is an indisputable duality about the individual. One aspect is distinctive individuality (or autonomy). The other is that of membership, of belonging, of being a part. The thing belonged to, the team, benefits from that individuality, but only when it is brought into play as part of the team, via membership. So that individual has to maintain individuality and at the same time surrender it via membership. Perhaps surrender is too strong a word, but for some team coaches it is hardly strong enough. Words like *sacrifice* come closer, indicating the primacy of the team over the individual. But if that sacrifice is wasted, a word not easily used, or at least when used never without attendant difficulty in the political arena, the team primacy is immediately suspected. Howsoever the belonging is expressed, it is very real and not without expense to that individual.

Let us also realize that there is not one single individual but many. This duality is replicated many times, with these instances being highly varied. It is through this variability that the system has its being. Not merely in the existence of members, nor in the distinctive roles that they play—helping the team toward dynamism—but in the diversity of expressions of this very duality of the maintenance and rendition of many distinctive individualities. It is as though this orchestra of autonomous beings surrendered yet kept their autonomy in mysterious unison. There is, if you will, a scale of diversity, of orchestration, of togetherness. At one extreme there is strict uniformity and no surprise, only increased volume. At the other there is a patternless cacophony. Somewhere in between these extremes the system comes into being, and performs well, when orchestration, however that is organized, is achieved. Diversity is more than difference, which is simply comparing one item with another, on some basis or other. Diversity is the measure of the paradox of the one and the many.

Here is how we express the paradox of diversity. First, we recall that a system is a collection of entities and their interrelationships assembled in such a way that the whole is greater than the sum of the parts. This notion of "greater than" has been summed up as "more is different."[13] The parts belong so as to serve the purpose of the whole. Yet this purpose is not well served if the parts belong for that reason alone. The homogeneity of partness is good, it ensures each and every individual is signed up as a part. But homogeneity is not something we want of the system itself. In order for it to survive and prosper, heterogeneity is required. How does the system inherit this heterogeneous quality? Our understanding is that this occurs when each and every part expresses simultaneously its individuality and its partness in manifold

Chapter nine: Paradox

and diverse ways. This diversity is what gives the system its heterogeneity. Thus, a tension is set up for each of the many between autonomy, maintaining individuality, and belonging, rendering this distinctiveness to serve the many. This aggregated tension creates the whole. The system is continually in tension, produced by this paradox of diversity.

9.4 Bothersome Bovines

We close this chapter by returning to our father's will, with which we opened. It is clear that this is an imperfect will. Only 17/18 of the herd is bequeathed. In other words, 1/18 of the herd, equal to 17/18 of one cow, is not available to the sons. For them to attempt to obtain this incomplete cow for themselves in whatever way would be to infringe an imperfect will. This raises questions. Is it wrong to infringe an imperfect will? Do two wrongs make a right? Was it the father's deliberate intention, his perfect will, to produce an imperfect will? And if so, what was his perfect will? And was this something he wanted his sons to discover?

Let us assume, not unreasonably, that the father knew what he was doing, that he had not made a simple arithmetical error, and that his imperfect will would be the means for his sons to discover something deeper.

Their first test comes with discovering the faulty arithmetic, which they accept. This gives them a dilemma. Do they try to settle their father's imperfect will perfectly? In other words, do they really try to obtain for themselves exactly 17/18 of the herd, no more and no less? To do so perfectly seems impossible without the shedding of blood, bovine not brother. The first son finds that his half produces eight cows, half a cow short of his exact entitlement. The middle boy gets five cows, 2/3 of a cow less than his proper inheritance. The youngest comes off worst. He has only one cow and is 8/9 short of what his father would have wanted, according to his imperfect will. With fourteen cows reasonably safely distributed, there are exactly three cows left. It must be tempting at this stage for the boys to reach a bloodless agreement and have one cow each. Let us look at the advantages. All of them gain. No cow is slain. The whole herd is dispersed. But on this last point, this represents an infringement of the father's will. Or does it?

The gain of each son is exactly in proportion to what the father would have wanted. The eldest boy gets an extra half cow (9 instead of 8½), the middle son gets an extra 1/3 of a cow (6 instead of 5 2/3), and the youngest boy gets an extra 1/9 of a cow (2 instead of 17/9). It is as if by being in agreement to be equal, instead of pursuing their proportionate inheritance, they meet their father's wishes. By disagreeing with their father they end up in agreement with him. Perhaps the imperfect will warranted this disagreement; perhaps this imperfect will inspired thoughts of equality, thoughts their father always had in spite of the proportioning his imperfect will contained. Perhaps his will is a conundrum? It contains unevenness and equality. It is fulfilled only by being violated.

But suppose the boys do not yield to this temptation—of dividing the remaining three cows equally. Suppose they pursue their father's proportions. The eldest boy should now get an extra 1½ cows. He takes one, giving him 9, and is owed a 1/2 cow. The middle son gets 1 cow. He now has 6 cows and is perfectly content. The youngest son has to be about as miserable as his older brother is deliriously happy. He gets no extra cows and is owed 1/3 of a cow. This apportionment also leaves 1/6 of a cow unbequeathed. Now suppose the elder and younger sons, who are as yet unable to have their rightful demands met, console one another. They are happy for their contented brother but they have to find a resolution to their unsettled business with their father. Suppose the eldest son now gets wisdom from above. He says to his youngest brother, "You have the half cow I am owed. And you take also the 1/6 of a cow that is unbequeathed. Put these together with the 1/3 that you are owed. What does that make?"

Of course it makes exactly one cow, the last of the herd of seventeen, which the youngest boy receives with the blessing of the oldest boy. Now they have what they would have had had they divided the three cows equally: 9, 6, and 2. The will has been violated again, but each boy comes out ahead as the father had intended, according to proportions, all the boys are happy, and the eldest boy has had the blessing of sharing with his youngest brother. Was this the father's perfect will?

Pursuing arithmetic made the neighbor redundant. But it turned blood brothers into good neighbors. There really never was an escape from this wisdom, just as there is never an escape from any paradox, without wisdom from above.

9.5 Time to Think

1. Consider all of the men in a small town as members of a set. Now imagine that a barber puts up a sign in his shop that reads "I shave all those men, and only those men who do not shave themselves." Obviously, we can further divide the set of men in this town into two sets: those who shave themselves and those who are shaved by the barber. To which set does the barber himself belong? The barber cannot shave himself, because he has said he shaves only those men who do not shave themselves. Further, he cannot not shave himself, because he shaves all men who do not shave themselves!

    Bertrand Russell, a philosopher/mathematician/political activist, who changed the direction of mathematics in the early twentieth century posited this paradox. It arises within set theory by considering the set of all sets that are not members of themselves. Such a set appears to be a member of itself if and only if it is not a member of itself. The significance of Russell's paradox can be seen once it is realized that, using classical logic, all sentences follow from a contradiction. In the eyes of many, it therefore appeared that no mathematical proof could be trusted once it was discovered that the logic and set theory apparently underlying all of mathematics was contradictory. By your own research, discover how this paradox led to breakthrough thinking in meta-languages. Discuss the implications of these findings for building ontologies and achieving semantic interoperability. Limit yourself to five thousand words.

2. You and a friend are presented with two boxes. The first box is transparent and contains $1,000. The other box is opaque and either contains nothing or $1 million. A mysterious benefactor offers you this choice and tells you that you may choose to take both boxes or just the opaque box.

"However," your generous benefactor cautions, "if I expected you to take both boxes, I have left the opaque box empty—you get only the $1,000." The mysterious person continues. "If I predicted that you would take only the opaque box, then I have placed $1 million in that box. You will get it all."

You and your friend begin to discuss what to do. Your friend wants to take just the opaque box. You argue that the benefactor has already made his prediction—the million dollars is either in the opaque box or it is not. It is not going to change. Whose argument is more correct? Discuss the implications of this paradox to the argument between determinism and free will. Limit yourself to five thousand words.

3. "Can an all-powerful being create something that is greater than itself?" is the central question of the omnipotence paradox. If a being is defined as being omnipotent, can it create a boulder that is too heavy for it to lift or a future that it cannot control? If it can, then it is not omnipotent, thereby violating our premise that the being is all-powerful; if it cannot, then it is again not omnipotent. This question is, of course, unanswerable, but its implications are still important. Consider its meaning in the context to making amendments to the Constitution upon which, in American jurisprudence, the supreme rule of law rests since the document is considered to be omnipotent. Specifically, "Can a constitutional amending clause amend itself—especially, can it do so when it is the only authority for the amendment, when it is the supreme rule of change in that legal system, when the new version of the clause would be inconsistent with the original, when the amendment would diminish the amending power, and when the amendment purports to be irrevocable?"[14]

4. A systems paradox that we chose not to elaborate upon in the chapter concerns that of in whom you believe, as a supplier of goods to a customer. It is stated as follows: You must listen to your customers in order to sustain your productivity. *But* you must also *not* listen to your customers in order to take advantage of disruptive technology. Refer to Clayton Christensen's work *The Innovator's Dilemma* and discuss this paradox in the spirit of the chapter, concluding with the breakthrough thinking that stems from releasing the tension of this paradox.

5. It is axiomatic that all of us are programmed to act in our best interests. Moreover, some believe that there is nothing we can do about this—the goal is set and we seek it irresistibly. This betokens who sets the goal though (and therefore what the goal is). If we set it ourselves, we could be wrong. If we let someone else set it for us, though, especially if he or she is omniscient and has our best interests at heart always, big ifs, then we cannot go wrong. Of course we have to believe, and we perhaps have to understand, the tension between determinism and self-will (see above). What is true for individuals might also be said to be true for corporations. What then is the best for them? Herein lies a fifth systems paradox: You must be upwardly progressive in order to increase profit margins, enlarge market share, and attack higher value markets. *But* you must also be downwardly visionary and mobile in order to temporarily live with lower profit margins, develop new markets, and establish capability in radically new technologies. Continue your researches into Clayton Christensen's work *The Innovator's Dilemma* and discuss this paradox in the spirit of the chapter, concluding with the breakthrough thinking that stems from releasing the tension of this paradox.

Endnotes

1. Taken from Titus 1:12–13.
2. Curly (played by Jack Palance) asks, "Do you know what the secret of life is?" Mitch, one of the city slickers played by Billy Crystal, waits for an answer. Curly holds up one finger and says, "This." Mitch responds curiously, "Your finger!" Curly retorts, "One thing. Just one thing. You stick to that and the rest don't mean shit." Mitch, confused but intrigued, asks, "But what is that one thing?" Curly, smiling, replies, "That's for you to find out."
3. For a person to cross a room he has to pass the halfway point, the quarter way point, and so on, ad infinitum. How can this infinity of steps accommodate a finite outcome, even though it is so evidently clear that a person can and does cross a room?
4. This reminds us of King Solomon's dilemma in determining which of two women was telling the truth as to who was the mother of a child. In the end he took, or at least proposed, executive action that led to a correct judgment, and if there is anything for which Solomon is better known than for his fabulous riches, it is his wisdom.
5. Quantum information processing; see www.qubit.org and http://cam.qubit.org/ and http://www.quantiki.org/wiki/index.php/What_is_Quantum_Computation%3F#What_are_qubits.3F.
6. Kaffee is assigned lead counsel by the JAG corps to defend two marines accused of the slaying of a fellow marine who ratted on one of them for an illegal fence-line shooting at their base in Guantanamo Bay, Cuba, where it is suspected "code reds" are used to discipline marines falling short of the required standards of conduct. Kaffee has to show that the two accused marines were following orders (that a code red had been issued by a commanding officer) and that the informant's death was accidental.

Chapter nine: Paradox

7. See, for example, http://www.wvi.com/~sr71webmaster/kelly1.htm and http://www.lockheedmartin.com/wms/findPage.do?dsp=fec&ci=16504&rsbci=15047&fti=0&ti=0&sc=400.
8. See, for example, http://library.thinkquest.org/27819/ch3_1.shtml.
9. Simplex is unidirectional communication duplex is bidirectional, and full duplex is simultaneous bidirectional. See, for example, http://en.wikipedia.org/wiki/Full_duplex.
10. "Go to the ant, you sluggard; consider its ways and be wise! It has no commander, no overseer or ruler" (Proverbs 6:6–7).
11. Dawkins, R., *The Selfish Gene*, Oxford University, London, 1976.
12. Lord Nelson and Sea Power, copyright 1995, 1998 by Leigh Kimmel, from http://www.geocities.com/Athens/3682/nelsonsea.html, reprinted with permission of Leigh Kimmel.
13. Anderson, P. W., "More Is Different," *Science*, 177, 393–96, 1972.
14. Suber, P., *The Paradox of Self-Amendment: A Study of Logic, Law, Omnipotence, and Change*, Peter Lang, New York, 1990.

chapter ten

Complex

10.1 Life's Rich Tapestry

Edward "The Confessor," so called because of his construction of Westminster Abbey, ruled England as her king for 23 years. Leaving no heirs, his death on January 5, 1066, ignited a three-way rivalry for the crown that culminated in the Battle of Hastings and the destruction of the Anglo-Saxon rule of England.

The leading pretender was Harold Godwinson, the second most powerful man in England and an advisor to Edward. Harold and Edward became brothers-in-law when the king married Harold's sister. Harold's powerful position, his relationship to Edward, and his esteem among his peers made him a logical successor to the throne. His claim was strengthened when the dying Edward supposedly uttered, "Into Harold's hands I commit my Kingdom." With this kingly endorsement, the Witan (the council of royal advisors) unanimously selected Harold as king. His coronation took place the same day as Edward's burial. With the placing of the crown on his head, Harold's troubles began.

Across the English Channel, William, Duke of Normandy, also laid claim to the English throne. William justified his claim through his blood relationship with Edward (they were distant cousins) and by stating that some years earlier Edward had designated him as his successor. To compound the issue, William asserted that the message in which Edward anointed him as the next king of England had been carried to him in 1064 by none other than Harold himself. In addition (according to William), Harold had sworn on the relics of a martyred saint that he would support William's right to the throne. From William's perspective, when Harold donned the crown he not only defied the wishes of Edward but had violated a sacred oath. He immediately prepared to invade England and destroy the upstart Harold. Harold's violation of his sacred oath enabled William to secure the support of the pope, who promptly ex-communicated Harold, consigning him and his supporters to an eternity in hell.

While the average English person would claim to know quite a lot about 1066 and all that, his or her knowledge is not often based on historical fact. The source of most of his or her information is the Bayeux Tapestry,[1] that colorful depiction of how William the Conqueror invaded England with his Norman army defeating Harold at the Battle of Hastings. But the tapestry is not a docile, dead depiction—it is alive with controversy and myth, providing us with a classic example of the old adage that history is written by the victors.

The tapestry is probably the most important pictorial image of the eleventh century. Arguably, it is one of the most important pieces of medieval art from any century. A work of enormous skill, it has priceless value as a piece of art in itself, and it is also an important source—a vital piece of historical evidence—for a key moment in Britain's national past. This does not, however, mean that its version of events is entirely accurate.

The tapestry was commissioned by William the Conqueror's half-brother, Bishop Odo of Bayeux, depicting the events surrounding the conquest. It details events leading up to the invasion and shows the key aspects of the conquest itself, not least the Battle of Hastings.

The tapestry is not a tapestry in the normal sense. It is actually an embroidery of at least eight colored wools, worked into pieces of linen. It is divided into a series of connected panels, approximately 20 inches wide and over 230 feet long. It is probably incomplete.

As edifying as this historical account might be, what does it have to do with the subject matter of our last chapter? The term *complex* derives from the Latin word *complexus*, meaning "to entwine." Its meaning today as an adjective is "consisting of interconnected or interwoven parts." As a noun it means "a whole composed of interconnected or interwoven parts." A complex really is a system. And being complex means being interconnected or interwoven, like, for example, a tapestry. The fact that the Bayeux Tapestry is not really a tapestry and is probably incomplete and inaccurate, in addition to being anomalous, adds spice to what we have to say about complex, because we now understand complex to be replete in paradox.

The science of complexity is relatively new. A scintillating account of its origins, historical and technical, is provided by Mitchell Waldrop.[2] A comparative newcomer to the science family, it is nevertheless possible to demarcate phases in our understanding of complexity. The initial phase was occupied by understanding the forces of self-organization, not realizing that self-organization was a system in its own right, for example, Adam Smith,[3,4] Friedrich Engels,[5] Charles Darwin,[6] and Alan Turing.[7] This was followed by seeing self-organization as a problem that transcended local disciplines and solving that problem, partially by comparing behavior in one area with that of another, for example, comparing slime molds and ant colonies. We might well consider to have now entered a new phase, which concerns the creation of self-organizing systems and the invention of artificial emergence: systems built with a conscious understanding of what emergence is, systems designed to exploit those laws, for example, software that makes book recommendations on Amazon, does voice recognition on our cell phone, and finds mates over the Internet.

The entire process of complexity understanding is driven by the paradoxical theme of unraveling that which is interwoven in order to understand the parts, their interactions and their interweaving, while keeping it in its

Chapter ten: Complex

whole state since the unraveling will fail to produce the extraordinary emergent behaviors attributable solely to the existence of that whole and to that whole's properties, which are meaningful only to it and remarkably different from properties attributable to the interwoven parts.

The paradoxes we have observed from our own studies and experiences of complexity include the following:

1. Complexity is much simpler than it first appears.
2. Simple things exhibit very complex behavior.
3. Little things mean a lot.
4. Myriad things are closer than we think.
5. Significant things are both vital and obscure.
6. Weak relationships bring strength and security.
7. To those who have, yet more shall be given, and yet to those who have little, even this will mean less.
8. A complex is both a one and a many, simultaneously (perhaps the ultimate paradox).

Our engineering friends talk of systems, subsystems, assemblies, piece-parts, systems of systems, families of systems, and complex systems. They leave complex to the end, as though all that goes before is somehow treatable differently and something special is reserved for real complexity. We think they might be missing a significant point. Then again, this is both vital and obscure.

Our goal is to better appreciate life's rich tapestry, and we believe that systems thinking plays a vital role in this. Life can be regarded not only as a search for meaning but a struggle to maintain order in the face of random disturbances. There are ongoing battles seemingly between chaos and order, between entropy and patterns, between simplicity and complexity, between the trivial and the enigmatic. These are examples of opposing forces locked in endless conflict. Instances of paradox. System is paradox. Systems thinking is our way of dealing with paradox, of respecting, valuing, anticipating, and leveraging complexity. And to do so is to appreciate an absolutely fundamental notion: the network *is* the system.

In *A Few Good Men*, Tom Cruise plays the part of an attorney, Lt. Daniel Kaffee, with a reputation for plea bargaining, a skill he has developed so that he can spend minimum time in the courtroom and maximum time playing and watching baseball. He gets a case to defend two marines, Lance Cpl. Harold Dawson and PFC Louden Downey, accused of murdering a fellow marine. Both defendants insist on calling Kaffee "Sir" lots of times, and once twice in one sentence. Kaffee has a line something like this, "You call me 'Sir' and I turn around and look for my father! I'm Daniel Kaffee, call me Danny." He wants to be their friend, not just their attorney, and the 'Sir' label is not helping. To the accused he is "Sir." But not to him. Not for the first time have we found terms misleading.

If you say "networks" to one of us (John Boardman), I turn around and think of copper wires and power lines, of solder, circuit boards, solenoids, and substations. I can't help it. It is all part of my inculcation as an electrical engineer.

You say "networks" to the other of us (Brian Sauser), and I turn around and think of the people I know, love, trust, and avoid. I think of social circles and of power networks of a different kind, of people in authority. I also think of the paradox between having leading-edge expertise gained through years of practice and involvement with others, an intimate belonging to networks, set against the need for independence when it comes to reviewing the work of others, of making an impartial judgment and setting aside subjectivity.[8]

That is the way we were made, the way we are. But not anymore. You say "networks" to us now and there is nothing that escapes our thinking. Electrical circuitry and social circles still figure. But to these we have joyously added the makeup of the human brain, food webs of our ecosystem and myriad other ecosystems, the structure of crucial proteins in our bodies, the grammar and structure of human languages, the Internet, and the World Wide Web. Talk about eclectic! It has been a struggle to leave behind our formative notion of the meaning of network, both physical and social. But it has also been a revelation because we are now in a far stronger position to realize the fundamental meaning of the term *architecture*. In fact, without this meaning tucked under our belt, we now realize the impossibility of understanding the meaning of the term *system*.

At the end of the movie, Dawson, having at one point called Kaffee a coward and told him he could not believe they let him wear the uniform, soberly and sincerely salutes him, recognizing and respecting a genuine authority gained by rapid maturity in the courtroom of all places. Kaffee still respects his father, but now he properly respects himself. Kaffee accepts the term *Sir*. We have come to accept what network truly means and we are about to share that with you.

10.2   Sides of Bacon

A Few Good Men was littered with stars including Jack Nicholson, Demi Moore, Kiefer Sutherland, and Kevin Pollack, in addition to the lead actor, Tom Cruise. The movie also starred Kevin Bacon after whom a curious game is named, the Oracle of Bacon at Virginia.[9] The point of the game is to determine for any actor their Bacon number. All of the actors who have appeared in a movie with Kevin Bacon, like Tom Cruise and Kevin Pollack, have a Bacon number of 1. Those actors who have not appeared in a movie with Bacon but who have appeared in a movie with an actor having a Bacon number of 1 receive a Bacon number of 2. And so on. An actor's Bacon number is an estimate of their his or her distance from Kevin Bacon, in terms of movie associations. To identify a Bacon number requires crawling over the IMD[10] database, which contains over 1 million actors and almost as many movies. We have done this in Figure 10.1. It is a big world and you might think that

Chapter ten: Complex

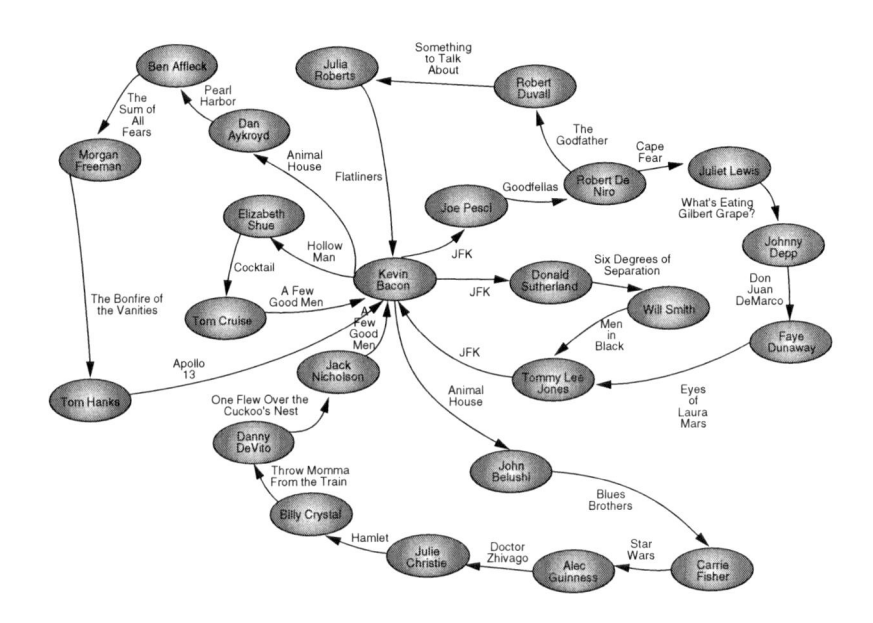

Figure 10.1 The oracle of Kevin Bacon.

the majority of actors lay some distance away from Kevin Bacon, unless in some strange sense he occupied the center of a compact universe. In fact, the opposite appears to be true. Less than ten thousand actors (1%) cannot be linked to Bacon. Even Pope John Paul II has a Bacon number of 3. The average Bacon number is less than 3. Fewer than 10% of all actors are more than 6 away from Kevin Bacon, 6 degrees of separation. As populous as the movie actor world is, it is evidently much smaller than it first appears. And not even Kevin Bacon has the lowest average degree of separation of all possible actors. That honor goes to Rod Steiger. Kevin Bacon's fame (which he never sought) as an epicenter for Hollywood is largely undeserved. There are almost one hundred actors nearer the center of this small world, including a good many we have never heard of!

The associations that actors have with one another via the movies in which they play in effect produce a network with actors appearing as nodes (or vertices) and links (or edges) being the movie(s) in which the two actors appear together. This type of network is a long way from the meshed distribution networks that provide electrical power to a small city. But network it is and properties it has, just like that massive physical engine that keeps a major city's lights ablaze. What is it that makes these two vastly differing entities similar? Is it merely abstraction, conceptualizing different objects by the same token, a node or a link? Or is there some substantive equivalence that integrates the two? What exactly is a network? And what is it about some networks that shrinks the world of objects that form that network? Is it true that on this planet of some 6 billion lives any one of them is separated by 6 handshakes (if not movies then some other form of association or acquaintance) from any other? How is this possible? And more fundamentally, what is the meaning of small?

10.3 A Weakness Stronger Than Strength

How many people do you know? Ten, fifty, one hundred, or more? Some of us know lots of people, others very few. In reality it is harder to really know many people very well. So some of our relationships, relatively few, are deep and regularly maintained. Others are more on the acquaintance level, and it is often difficult or too bothersome to keep these latter relations going. Over time they usually whither and die. Their demise at least can be considered at the expense of strengthening the few that really matter, however those are decided.

Some of your circle will know people you do not know, which in a way extends your network. But most of the people you know will know one another. These circles are probably better referred to as a cluster, a reasonably tight knitting together of a close group of friends. Clusters make up a world, but they do nothing to make it small. Strong ties hold a cluster together, but it is the weak ties that turn a collection of clusters into a small world.[11]

It is this paradox, that weak ties are what gives a small world its strength, that for us typifies complexity and systems thinking. It is one of many, as we shall discuss. But lingering with this a little is worthwhile, for we must continue to break out of a mindset that insists a network is either a circle of friends or a circuit of transformers, power lines, and switch gear. Weak ties explain the computing power of our human brains and the synchronicity of fireflies in a tropical rainforest in Papua New Guinea. From friends and fuses to fireflies and firing neurons. We are beginning to explore the groundbreaking science of networks, and with it immense opportunities for complexity understanding and systems thinking.

The articulation of a small-world architecture using the mathematics of graph theory is comprehensively captured by Duncan Watts.[12] A paper by Watts and his thesis advisor, Steve Strogatz, largely free of mathematics, set ablaze huge interest in the phenomenon of small worldness, with its architectural fingerprints being found in diverse fields as ecosystems, natural language, and the World Wide Web.[13]

At the outset of their work together they sought to introduce random links between a fully ordered network of clusters. Figure 10.2 shows this increasing randomness from regular networks to random networks. Suppose the initial circle (of friends, say) has one thousand dots, each connected to its ten nearest neighbors. This gives in all about five thousand links. To this let us add ten links at random (0.2%). The network is still essentially ordered but is now lightly splashed with randomness. More and more links can be added gradually at random, and the effect on clustering and on any perceived small worldness calculated. This is a form of network evolution with

Chapter ten: Complex 193

Regular Network   **Small World Network**   **Random Network**

P = 0   Increasing Randomness   P = 1

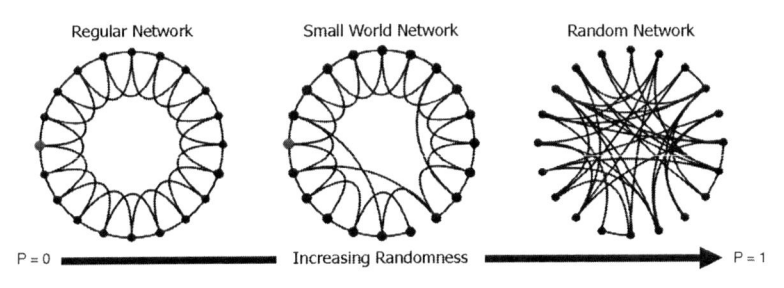

Figure 10.2 Increasing randomness.

strict order being updated with random rewiring. They found that while small disturbances had no noticeable effect on clustering, they had a devastating effect on small worldness. Initially the degrees of separation was 50, but with a few random links it plummeted to 7. Skeptical of their findings their experimentation continued with greater scrutiny. No matter what they did, however, they always found that the lightest dusting of the ordered network with random links produced a small world.

In a planet of 6 billion, with each person linked to their 50 nearest neighbors, the number of degrees of separation is of the order of 60 million. Throw in a few random links, a fraction of the total being 0.02%, and the degree of separation drops to 8. If the fraction is slightly increased to 0.03%, it falls to 5. Clustering, a social reality, persists, but small worldness, a counterintuitive phenomenon, appears. Courtesy of random encounters, forming relatively weak relations, but strong enough to tie a planet together. Little things do mean a lot! Now what about those fireflies and fiery neurons?

10.4 Ready, Fire, Aim

Imagine a 200-yard stretch of forest bordering a river in Papua, New Guinea. The trees are 40 feet tall. The scene while verdant and panoramic is nothing extraordinary. A firefly decks each leaf, but you see none. Night falls. Speckles of light dapple the stretch and interest awakens. Soon there are clusters of blinking lights as near neighbors get accustomed to their fellow flashers. The scene crescendos in a series of single solid flashes, about twice per second, along the entire stretch. Millions of fireflies have synchronized themselves into an orchestra of light in the darkness. It is a vista to rival anything that Walt Disney pulls off at Epcot. The scientific term is terrestrial bioluminescence.[14] But let us not allow technical nomenclature to obscure an inexplicable phenomenon.

How is this possible? How is synchronicity achieved? Is there a conductor for this orchestra? The ant, we are told, has no commander. Are fireflies somehow a brighter species, that a leader emerges from their uniform ranks? And if so, is it possible for millions of fireflies to notice a single leader and be smart enough to subordinate themselves to this single command? Perhaps not. By the same token it appears unlikely that fireflies have the bandwidth to tune in to all their neighbors. If there were, say, ten thousand fireflies, the total number of communication paths between them would be 50 million. Can a single firefly monitor ten thousand chat lines in the context of 50 million traffic lanes? Unlikely. The motivation is high. The flashing is the means by which males attract females as a prior to mating. It makes sense to produce a series of single blinding flashes. That will get the females' attention. Going it alone is risky. But can a forest of fireflies produce the collective consciousness to synchronize in order to maximize mating potential? Who has that idea? Some of them or all? And how do they share that notion?

To make progress with these questions we need again to turn to the seminal work of Duncan Watts and Steve Strogatz. It seems possible that near neighbors will somehow get their flash act together. By comparison, two grandfather clocks near to one another in a room have exhibited a synchronism of their pendulums, the explanation being the interaction of rhythms each transmits to the other via the floor. Let us argue then that each firefly responds mostly to the flashing of a few of its nearest neighbors. The computational burden on a firefly is now more realistic, tuning in to, say, five neighbors. The traffic lanes are now reduced to 0.1% of what they were when each firefly could communicate with any other of a population of ten thousand.

A rare few fireflies might also feel the influence of a fly or two at a longer distance. A few fireflies might have a particularly brilliant flash, and so be visible to others far away, or few genetic oddballs might respond more to fainter flashes than to bright ones. In either case, some fireflies make it possible for long-distant links to exist between evident clusters. This argument begins to present the opportunity of a small-world pattern.

Watts and Strogatz carried out computer simulation using this small-world architecture and repeatedly found that the insects were able to synchronize almost as readily as if each one had the power to speak to any other fly. In essence, the pattern is a breakthrough in computational efficiency.

No one really knows how fireflies are connected. Only a few species in Malaysia, New Guinea, Borneo, and Thailand have the power to synchronize evidently. So the terrestrial light orchestra is still shrouded in mystery. But a fingerprint of computational speed and power may have emerged. Armed with that, it does not seem unreasonable to ask whether our human brains, computation engines par excellence, might possess a small-world architecture.

Phrenology is thoroughly discredited as a body of knowledge. But it was not entirely without purpose. The notion that the brain is somehow arranged into organs, functional building blocks each devoted to specific tasks such as memory, sight, sound, emotion, and so on, is one that has usefully carried over from the hocus-pocus of feeling the lumps and bumps on our skull as a pattern match with personality traits, to modern-day neurological science.

The cerebral cortex is held to be the locus of our higher capabilities. This thin, gray, intricately folded and delicately packed outer layer of the brain,

Chapter ten: Complex

just a few millimeters thick, contains the precious neurons from among the 100 billion that make up the brain's tissue. The cortex is the part of the brain that lets us speak, perform mathematics, learn music, and invent excuses for being delinquent. It is what makes us distinctively human. And it is indeed organized into something like a set of organs. MRI scans are ways of detecting neural activity based on oxygen content in the blood flow patterns around the brain. They can therefore act as a lens into the modular decomposition of the brain relative to various tasks in which we are engaged, for example, responding to a verbal command, recognizing a taste, or recalling a friend's address.

Squeezing 100 billion neurons into a 3-pound lump inside the skull seems far fetched, but not when you consider that you can get more neurons into a thimble than there are people in the United States. Crudely, a neuron is a single cell with a central body from which issue numerous fibers. The shortest of these, called dendrites, are the cell's receiving channels. Longer fibers known as axons are the transmission lines for the neuron. Most of the neurons link up with near neighbors within the same functional region. Signals from axons are received by dendrites, and so neural activity is myriad signaling along these channels, of which there are hundreds of trillions. The brain is interwoven like nothing else we know. It is complex. But is it simpler than it seems?

While functional regions are in effect clusters of neural connections, the brain also has a smaller number of truly long-distant axons that link brain regions that lie far apart, sometimes even on opposite sides of the brain. Consequently, the human brain has many local links and a few long-distance links, something that starts to resemble a small-world pattern. Research has shown that the degrees of separation in a cat's brain is between 2 and 3, identical with a macaque brain, while at the same time regions are highly clustered. So it seems that what is true of social networks is also true of what Mark Buchanan charmingly calls a thoughtful architecture.[15]

Some biological advantages of this small-world architecture in the human brain are compellingly clear. If you accidentally hit your thumb with a 5-pound hammer instead of the intended nail, several things happen in coordinated fashion. You drop the hammer, draw the offended thumb to your mouth and suck it, let out a scream, and do a jig. At least that is what we do! Several parts of the brain are called upon to engage this kind of bodily function, and that orchestration is achievable only because of the long-distance links that tie the various clusters responsible for separate actions together. A second advantage lies in the fact that brain clusters provide huge redundancy, so with the wear-out or fallout of neurons, the functional blocks can still perform their functions. Even if functions are degraded or rendered impotent, their separation means each block can still function so that loss of speech understanding does not necessarily mean loss of memory or the ability to make future plans. Even if communication links are broken between, say, the visual cortex and the hippocampus, which could result in a slight degradation in short-term memory of visual information, the small-world architecture takes care of that by providing alternate, longer, less direct routes. It is as if people remain neighbors even though gulfs well up; they simply use go-betweens. After all, these neurons are all in it together, all 100 billion of them!

There is one final thing to say about the brain's magic imparted by this thoughtful architecture. After construction and coordination comes consciousness. No one knows where this resides. It is one thing for neurons to be ready and to fire—they deserve their 5-millisecond reset time having unloaded—it is another thing to take aim, to say, "I am a conscious human being, I am me, and I am unique." How does this come about? The orchestration of billions of neurons might be addressable in terms of small-world architecture and synchronicity, but how does this self-organization at the *many* level lead to self-awareness at the *one* level? State-of-the-art research seems to suggest that there are connections between the two, and it is all a matter of the simultaneity of the many and the one, the ultimate paradox of complexity.

Thinking of consciousness requires us to address two aspects: conscious states and conscious organization. Consider the scenario of being a student in a classroom and briefly, while being unengaged by your instructor, you glance through a window to see someone running toward your building. What do you make of this? More particularly, what does the brain make of this? It engages in the generation of multiple states: pattern recognition, movement detection, context setting, generation of emotions, awareness of sounds, selection of possibilities. All at the same time and all concluding in a single integrated picture.

This is made possible by neural synchrony: the coherent engagement of neurons in many regions of the brain and *at multiple levels* into one overall pattern. Research has found that when the brain is confronted by two distinct views, of some simple patterns, neural activity is not synchronized when the patterns are seen separately, but when they are made to merge and become one pattern, neural activity is synchronized. It is synchrony that creates conscious integration. Moreover, in synchronous movement individual neurons maintain a subtle but defined lead or lag behind the group's average firing so that the whole orchestration is information rich in what it provides to upper-echelon neural circuitry. It is the equivalent of hearing the orchestra and the violins and the flute. The one and the many.

One of our paradoxes in coming to an understanding of the term *complex* is that complexity is far simpler than it looks. Behind the apparent confusion lies a hidden order. So this proves in the small-world architectures of Watts and Strogatz. That order can be summarized by tightly knitted clusters and random connections between these. It is as if the randomness gave rise to the order we find in the computational efficiency and resilience that instances of these architectures produce, for example, finding female fireflies and mind magic. This is encouraging. We turned the paradox to our advantage. Let us not leave it there. What else? Is it possible that order lurks behind pure chaos? Are there invisible forces at work to shape our lives, our technologies, and our environment regardless of happenstance, uncertainty, and accident? Can there exist a design presence that steers a course while chance itself holds sway? This is not merely an academic question or idle philosophy on our parts, though we do take great enjoyment in posing these questions for their own sake.

There are many instances of networks that exist in the real world that simply do not conform to or are shaped by the forces that give rise to the small-world architectures we have thus far enunciated. Inspired as we were to abandon our stereotypical networks of electrical grids and social circles to go on and discover new varieties of this key notion, the firefly flocks and neural networks being prime examples, we can find many more examples— river networks, air transport networks, the Internet, and the chemistry from which we humans are made. None of these exhibit the preexisting clustering that our small-world examples have required. These are all products of two forces: growth and chance. Who knows when it is going to rain, how heavily, and where? Water that fills our rivers. Who decides what airports there shall be served by what kinds of aircraft traveling to who knows where? Planes that fill our skies. What determines which servers will attach to the Internet, publishing material to the Web and providing access to countless millions? Servers that thrill our surfers. As much as we might imagine we have an involvement in these things, in no way can we say we determine outcomes. These are governed by growth and chance. Do these dance? And is there a discernible choreography?

10.5 *Snowballs and Seesaws*

What do the great Mississippi, the Internet, and a computer game inelegantly named diffusion limited aggregation (DLA) have in common? At face value the answer is surely nothing. But that is because we are looking for a simple explanation. To find what we are looking for we must subject ourselves to paradox, and find the complexity in each of these systems.

The Mississippi River stretches 2,315 miles from its source at Lake Itasca in the Minnesota North Woods, through the mid-continental United States, the Gulf of Mexico Coastal Plain, and its subtropical Louisiana Delta. Its river basin, or watershed, extending from the Allegheny Mountains to the Rocky Mountains, including all or parts of thirty-one states and two Canadian provinces, measures 1.81 million square miles and covers about 40% of the United States and about one-eighth of North America. Of the world's rivers, the Mississippi ranks third in length, second in watershed area, and fifth in average discharge. That is the big picture. How it got formed is largely by accident.

Over however many years, the clouds gathered, the rains came, the ground washed away, and the mighty Mississippi began to take its shape. We do know gravity played a part; that is why the rain comes down! As the soil erodes by the washing of rainwater, channels are formed. This has a positive feedback effect in enhancing that flow of water. Grid lines are carved on the earth as fingerprints of the rain's reign, the watermark of myriad deluges. An invisible force is at work influenced by gravity and history that of what has gone well before will be welcomed back again. What is the imprint of this force? To answer that question we need an imaginative leap into data capture. What is the relationship between sectional lengths of the river and the amounts of water these drain? Why should we bother to ask this question? The reason for that is because the imprint of the Mississippi bears marked similarities with many other great river basins. There exists a pattern in the formation of river systems, and this cannot be explained by comparing the details of their environs or history (see Figure 10.3). But it can be accessed by this imaginative question.

These data conform to what mathematicians call a power law, or what engineers call log-log. And it is this power law relationship that holds true for all river systems. It becomes attractive to think of this as the architecture for river systems, a design whose architect eludes us without faith.

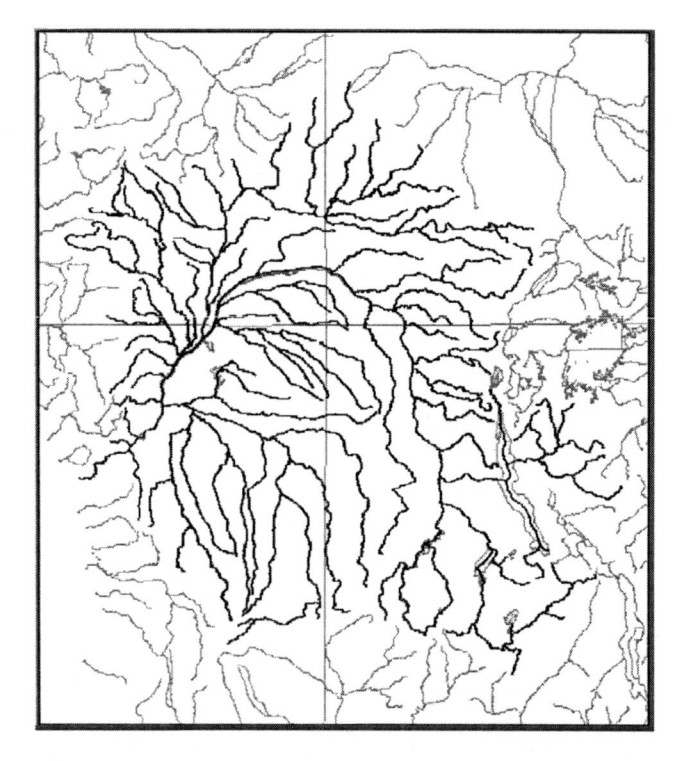

Figure 10.3 River drainage. (From Maidment, D. R., "Creating a River Network from the Arcs in the Digital Chart of the World," Kwabena Asante and David Maidment, Center for Research in Water Resources, University of Texas at Austin, November 1999. Reprinted with permission of David R. Maidment.)

We cannot improve upon Mark Buchanan's eloquence: "The real importance of the power law is that it reveals how, even in a historical process influenced only by random chance, law-like patterns can still emerge. In terms of this self-similar nature all river networks are alike. History and chance are fully compatible with the existence of law-like order and pattern."[16]

Does this power law have ubiquity? Can it be that this architecture shows up in the Internet and this clumsily named computer game?

The Internet has its origins in many ideas, including a DoD need to build resilient infrastructures, as well as the need of many researchers to share information reliably and efficiently, initially confined to California. Its origins are, however, of far less importance than its history, as remarkably brief as this is. That history has been governed by spectacular growth and inordinate randomness. No one determines the Internet, even though its protocols are well-published standards that are fully obeyed. Is a pattern possible for such a system? The relationship between the number of nodes in the Internet and the number of links these nodes possess to other nodes (indicated by the number of routers located at these nodes) (see Figure 10.4). Once again this relationship conforms to a power law—it is log-log.

Apparently rain falls where it will, causing rivers to flow where they will, and routers sprout wherever they will, creating information flows where they will. But in both cases they will be done according to a higher power law.

What about the DLA game? Imagine a blank screen and an insignificant anonymous object drifts across. A second one does likewise, both appearing and traveling perfectly at random. If they bump into one another, they stick together. If they miss, they carry on their random walks, perhaps disappearing from view. Millions of these objects appear over time. A figure appears. It should, according to our intuition, be an anonymous insignificant blob—an aggregation of myriad identical objects. But there is a pattern (Figure 10.5). It is tentacled, from which we infer that it is hard for new objects to get to the center of this nonblob. The tentacles are self-reinforcing. More than this, they are self-similar. What mathematicians call fractal. Just like the river systems. Our power law just will not go away! Growth and chance keep it alive and well. It shows up in the most surprising places. In our body chemistry, one or two specific molecules take part in several hundred chemical reactions involved in the bacterium's metabolism, whereas many thousands of other molecules take part in only one or two reactions. The distribution of molecular interactivity against the number of molecules with a given interactivity is yet another power law.

Given its irrepressible nature, might we dare ask: Does the power law provide us with a network signature? What is this like? And what rules are in operation that create or govern these power networks?

Though a little naïve, it is hard not to equate the power law with "to the victors the spoils." That is how power operates, right? A colloquial expression for this is "The rich get richer." As much as one might despise this, it seems

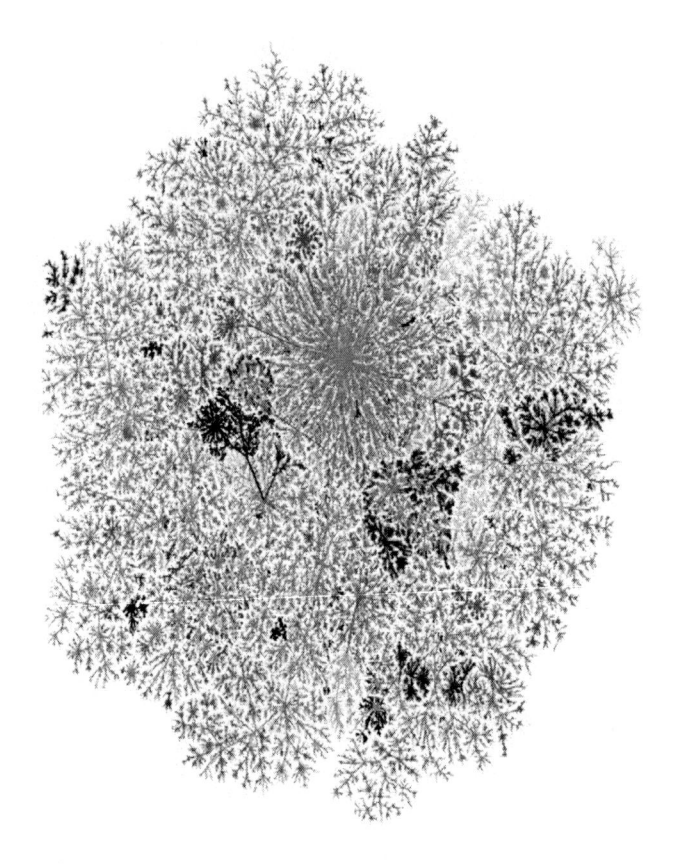

Figure 10.4 The Internet (graphical depiction).

to be a law of the universe. Is it the power law? Barabasi and his colleagues[17] answer this for us by their experiments with preferential attachment.

Consider a green field situation—several nodes and no links. Gradually links are added entirely at random. With some new links come new nodes as they enter the growing network. Inevitably a few nodes will gather a few more links than others. Now consider that as new links are added they have a preference toward connecting to nodes that already show a preponderance of links to other nodes. The process of adding links is still random, but the probability that these now attach to the more popular nodes is slightly increased. With such a set of rules these experiments produced a similar pattern repeatedly. What is more, the architecture of these patterns always conformed to a power law, with a few nodes acting as powerful hubs and myriad nodes having relatively few links. The number of nodes plotting against the population of links that these nodes support falls off in log-log fashion.

Chapter ten: Complex　　　　　　　　　　　　　　　　　　　　　　　　201

Figure 10.5　Diffusion limited aggregation.

We observe this phenomenon in many walks of life and instances of science. A power law fits the number of nonexecutive company directors with a precious few holding more than one hundred offices and very many just one or two directorships. It is clear why. Corporations need savvy to inform their strategic planning. So much of this can come from nonexecutive directors. The ones that are most coveted are the ones that are already popular with companies, that is, who are already serving in many capacities. It makes sense. Each corporation gains via the wildly popular nonexecutive director the enhanced experience that person is gaining courtesy of serving on several boards already. These folk are in effect conduits of corporate knowledge around the landscape. Conduits embodied as hubs. It is the old boys' club writ large.

We see the power law in sexual-contact networks. Inevitably a few people are more sexually active than others in terms of the partners they have. There are forces at work here. With success at gaining new partners comes an acquired skill to gain yet more. With more partners gained comes the need to practice that skill more extensively in order to keep up a good image. With that motivation comes more skill and more partners. It is a cycle that Pete Senge would call a reinforcing loop, characterized by a snowball rolling down the side of a mountain potentially producing an avalanche. A hit song from Queen[18] could not be more apposite to capture this momentum:

> I'm a rocket ship on my way to Mars
> On a collision course
> I am a satellite
> I'm out of control
> I'm a sex machine ready to reload
> Like an atom bomb about to oh oh oh oh oh explode!

This type of network also has the small-world property. It is another flavor of small that causes us to differentiate between the Watts and Strogatz variety and that of Barabasi and colleagues. The former has been termed egalitarian and the latter aristocratic. Examples of the nouveau riche are the in-demand nonexecutive directors and the sexually prolific. A more obvious example is the wealthy themselves (Figure 10.6).[19] Here is quite literally a case of the rich getting richer. Money flows between people are essentially transaction based: You give me work, I give you money. You give me money, I give you goods. This works for all of us alike. But money in return for time or goods is very limiting. It is an activity that characterizes egalitarianism. The real power comes when money works for you.[20] This involves risks but carries rewards. Big risks carry huge rewards. Two things characterize rich thinkers. First, they put less value on their money because they have so much of it. Risks are reduced accordingly. But risks are further reduced by focusing the time that is not spent working for money on being smarter about what will work and what will not. Investment to the rich does not equate with gambling by the poor. There will always be risks since uncertainty rules, but you can minimize these by investing your time wisely.

Is there an end in sight to these gains? Does reinforcement continue endlessly? Or are there limits to growth as Peter Senge found. Does the snowball meet any obstacles that can prevent the avalanche? Is a seesaw in sight?

Left unchecked, the air transportation network in the United States would become aristocratic by nature, with the major hubs being, of course, Atlanta,

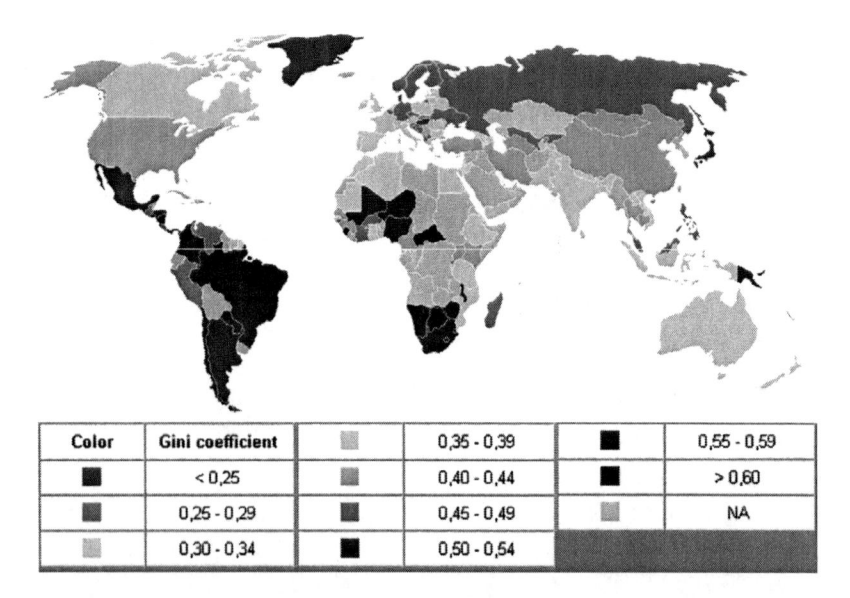

Color	Gini coefficient		0,35 - 0,39		0,55 - 0,59
■	< 0,25		0,40 - 0,44	■	> 0,60
■	0,25 - 0,29	■	0,45 - 0,49		NA
	0,30 - 0,34	■	0,50 - 0,54		

Figure 10.6 Wealth distribution: world map Gini coefficient.

Chicago, and Dallas. The hub-and-spoke system much favored of airlines serves their needs to carry as many passengers as possible wherever they want to go, provided it is via their hubs. It makes economic sense for them to concentrate resources and facilities at major airports at the inconvenience to passengers of switching flights and layovers. But the 7 million passengers that pass through Atlanta's Hartsfield–Jackson International Airport annually represent a limit to growth. The airport is often running beyond maximum capacity, and when bad weather shows up, not only in Atlanta but at connecting cities, life gets hairy. Passengers vote with their feet, which explains the growth in regional airports, smaller aircraft, and point-to-point travel. People are not electrons. There is a distinct difference between atoms and bits.[21] And whereas there appears no limit to how many Web sites can point to and be pointed at by others, this is not the case for mere mortals with luggage and a persistent need for burgers, bathrooms, and beds.

So there are balancing forces that will arise to keep in check the "rich gets richer" snowball. The interesting thing for us to consider is how these network architectures, aristocratic and egalitarian, can switch. What are the factors that determine this, and what are the consequences for people, corporations, species, and technology systems in terms of reliability, security, safety, and resilience?

10.6 Significant Others

The persistence of the small-world architecture is impressive. That it comes in these two flavors is also charming. Both types emphasize the paradox of revealing a hidden order to apparent chaos and of providing a simple elegance to what otherwise seems immensely complex. Who can be satisfied, though, with leaving matters there when the urge to find deeper meaning through higher-order patterns has been stimulated by successes thus far? Isn't it the case that we are in a process of understanding significance, of meaning itself?

In the egalitarian networks the significance lies in the weak links, another extraordinary paradox. In the aristocratic network, more evidently the significance lies in the (super) hubs, also known as the vital few (compared to the trivial many). What exactly is the nature of this significance? That, of course, depends on the real-world situations whose apparent disorder and complexity are elegantly captured by small-world architectures.

In the case of the world's ecosystem, aristocratic small-world structure is a natural source of security and stability. Yet the super hubs or "keystone species" represent crucial organisms the removal of which might bring the web of life tumbling down like a pack of cards. Removing even 20% of the most highly connected species fragments the food web almost entirely, splintering it into many tiny pieces, doing untold permanent damage to the web of life. Culling of one species sends out "fingers of influence" that in a few steps touch every last species in the global ecosystem. Strong links between species set up the possibility of dangerous fluctuations therefore, since the vital few are the vulnerable feet of an aristocratic giant. By sharp contrast, the weak links between species act as natural pressure valves in communities. The weak had once again gone unnoticed since our concentration was focused on the vital few hubs. Paradox demands wisdom, and complexity often finds us lacking in that department.

A smart David spots the temple of his opponent. An agile youth unencumbered by heavy protective armor for which no need is foreseen casts the first stone. And it is enough.[22] The lesson for us is to find the simplicity behind the complexity, recognize what is significant and, in the case of food chains, be smarter about what we can and cannot cull.

Resilience, the ability to withstand major disturbances and quickly restore order, is now a matter of architecture in the face of specific threats. Random networks, despite their redundancy, fall apart quite quickly in the face of an uncoordinated attack. The aristocratic network, like the real-world Internet, falls apart gracefully under random attack and does not suffer catastrophic disintegration. But the very feature that makes an aristocratic network safe from random failure could be its Achilles' heel in the face of an intelligent assault. As far as the Internet's wholeness or integrity is concerned, the destruction of 18% of the most highly connected hub computers serves to splinter the network almost entirely into a collection of tiny fragments.

10.7 Postlude

We cannot close this chapter without recognizing that we are also closing this book. As Looney Tunes[23] had it: "That's all folks!"

Our challenge then is to draw a curtain on our present subject matter—complexity—and at the same time conclude our descriptions as a whole—*Systems Thinking: Coping with 21st-Century Problems*. This is apposite because if there is one thing that characterizes these problems, it surely is complexity. It seems that no matter what we do about our problems (or how we do it), the solutions we conceive and develop always lead to greater difficulty. At best, they stave off the evil day. At worst, they make matters worse. Considerably so.

Maybe C. West Churchman[24] had it right—we are locked into an endless cycle of deception and perception. With deception comes perception (a solution), but with that comes more deception (another problem). Then we see it is no problem at all (perception) and do something clever about it, only to be disappointed (deception). Not just with our new failed state, but also with our failed thinking. The alleged process that took us from one state to the other. Is there any way out of this?

Peter Checkland realized that there was a "problem of problem definition" and gave us Soft Systems Methodology. Peter Senge and many others realized that what appears simple is actually very complex and gave us a tool kit to articulate that complexity and thus simplify it. Others have shown us, emphasized in this chapter, that that complexity is simpler than it first appears. The trick is to find the simple expressions that capture that complexity and unravel it for us.

People like us are largely observers and commentators. Nevertheless, as potential originators, we have offered you our own toolkits: the conceptagon, the systemigram, and chiefly our take on the meaning of *of* and the value of paradox. What we have tried to draw your attention to is the use of these tools to help determine the real meaning of significance. In a systemigram we try to establish the significant elements, nodes, and links that together form a whole not only greater than but different from the parts. In other words, to emphasize holistic thinking and emergence. In our treatment of *of* we try to establish the significant architectures that draw whole systems together in order to produce not simply emergence but an emergent culture, one that relies on emergence, the element of surprise, the risk of chaos, in order to produce new patterns of simplicity and elegance about which we could not know or foresee. In our treatment of paradox we try to reestablish the significance of wisdom and where it might be found. In the multiplicity and simultaneity of viewpoints. In respect for competing perspectives. In the duality of the one and the many. In the conflict and the confusion that arises when we refuse to accept simple choices and with wisdom find a simplicity unforeseeable by mere resolution—something that is hidden and can only be found in the unthinkable.

And we have not sought to do these things merely for the sheer joy of it, though that to us is of vital importance. We have done so to be practical as our constant references and illustrations hopefully show. Perhaps we should end where we began. In Iraq. Or out of Iraq? It was an ever so simple question. But we always knew it to be hiding immense complexity. Beyond this, does a more natural simplicity lie, accessible only to wisdom? What are the threads of this complexity? The oil? The love of freedom? Our need for world peace? Democracy? The American way of life? The evil of empire? The good of mankind? And how are these interwoven?

Our world is small. Are all men created equal? Is that the nature of our smallness? Or do empires hold the vital keys to our smallness? Of either Islamic fundamentalism or agnostic capitalism? If the former, then the weak links are significant. If the latter, the vulnerability of empires makes the weak links vital. Is wisdom drawing us to find strength in weakness? And do we have the strength to go there? What do you think?

10.8 *Time to Think*

1. In his book *Nexus*, Mark Buchanan writes:

   > Physicists in particular have entered into a new stage of their science and have come to realize that physics is not only about physics anymore, about liquids, gases, electromagnetic fields, and physical stuff in all its forms. At a deeper level, physics is

really about organization—it is an exploration of the laws of pure form.[25]

    Is this physics overstepping the mark, being imperialistic, and wanting to be the aristocrats if not the monarch of all science? Or is Buchanan saying something radically new, that matter is not merely stuff but an organization, an arrangement, kinds of togetherness? Is he saying the network is the stuff? If this is so, does this make it more meaningful to speak of a biology of systems whereby that science's knowledge becomes less about the stuff it has explored and more about it being a metaphor for systems, which are also conventionally regarded as stuff, as really being architectures, networks, and pure form? Discuss.
2. Weak links, or relatively little exercised relationships in a social context, are what make a huge world small and what give that small world its strength. By contrast, the strongly connected hubs in an aristocratic network are also its weak points since their targeted removal shatters that network. Asymmetric threat is a phrase that conjures a relatively weak adversary whose impact nonetheless is out of all proportion to its strength, while incomparable military might appears weak. Is this just a play on words that we can safely ignore? Or does this juxtaposition of weakness and strength force us to look for deeper meanings? And is part of that search a need to reexamine the way we think?
3. One cannot have a conversation with a single neuron, and yet in some sense neurons can communicate with one another, and as a result 100 billion of them enable each of us, as individuals, to have a conversation with one another. Neuron chatter emerges as intelligible speech. By comparison, notwithstanding the BBC's motto "Nation shall speak peace unto nation,"[26] individual chatter somehow does not translate into intelligent crowd speak. What are the lessons of collective consciousness and computational efficiency exhibited by what Buchanan describes as "the thoughtful architecture," that is, the brain, for us to learn in terms of nation-states and the ways in which they exchange ideas? Do Thomas Friedman's ideas[27] give us direction here?
4. Limiting the spreading of a disease, for example, HIV, is a matter not only of medicine but also of education. But just as the design of drugs is specifically targeted at the biochemistry of the disease, so also must the education be focused with equal precision. We take aim according to the architecture of the social network that transmits the sexual activity. If this is of the aristocratic kind, the super hubs are our bull's-eyes. If it be egalitarian, the weak links, being what makes this world small and vulnerable to epidemics, are the target. Identification of these nodes and links is nontrivial, although of equal significance to that of medical research. Here is an example of the equality of stuff and pure form in the articulation of remedies. In regard to the spreading of news, good or bad, select a topic of your choosing (e.g., in the case of good news,

how to get to heaven after you die, or if you are an illegal immigrant, how to get into the United States; and in the case of bad news, gossiping or enemy propaganda) and then describe how stuff interacts with the organization of form. Conclude your description with an agenda for change, that is, improvement or remedy.

Endnotes

1. See http://www.bayeuxtapestry.org.uk/.
2. Waldrop, M. M., *Complexity: The Emerging Science at the Edge of Order and Chaos*, Simon & Schuster, New York, 1992.
3. Smith, A., *The Theory of Moral Sentiments*, Access from the Library of Economics and Liberty, 1759.
4. Smith, A., *An Inquiry into the Nature and Causes of the Wealth of Nations*, Access from the MetaLibri Digital Library, 1776.
5. Engels, F., *The Origin of the Family, Private Property, and the State*, Hottingen-Zurich, 1884.
6. Darwin, C., *On the Origin of Species by Means of Natural Selection; or the Preservation of Favoured Races in the Struggle for Life*, John Murray, London, 1859.
7. Turing, A. M., "Computing Machinery and Intelligence," *Mind*, 59, 433–60, 1950.
8. Sauser, B. J., "Attributes of Independent Project Reviews in NASA," *Eng. Manag. J.*, 18, 11–18, 2006.
9. See http://oracleofbacon.org/how.html.
10. Interactive Movies Database, www.imdb.com.
11. Granovetter, M., "The Strength of Weak Ties," *Amer. J. Soc.*, 78, 1360–80, 1973.
12. Watts, D., *Six Degrees*, Vintage, New York, 2004.
13. Watts, D. J., and S. H. Strogatz, "Collective Dynamics of Small World Networks," *Nature*, 393, 440–42, 1998.
14. See, for example, http://www.dartmouth.edu/~dujs/2000S/06-Biolumen.pdf.
15. Buchanan, M., *Nexus*, Norton, New York, 2002, 64.
16. Ibid., 103.
17. Barabasi, A.-L., *Linked*, Penguin, New York, 2003.
18. *Don't Stop Me Now* is a 1979 hit single by Queen, from their 1978 album *Jazz*. Words and music were by Freddie Mercury.
19. The Gini coefficient is often used to measure income inequality. Here, 0 corresponds to perfect income quality (i.e., everyone has the same income) and 1 corresponds to perfect income inequality (i.e., one person has all the income, while everyone else has zero income).
20. Kiyosaki, R. T., and L. L. Sharon, *Rich Dad, Poor Dad: What the Rich Teach Their Kids about Money—That the Poor and Middle Class Do Not!* Business Plus, New York, 2000.
21. Negroponte, N., *Being Digital*, Vintage, New York, 1996.
22. 1 Samuel 17.
23. See http://en.wikipedia.org/wiki/Looney_Tunes.
24. Charles West Churchman was an American philosopher in the field of management science, operations research, and systems theory. He is internationally known for his pioneering work in operations research and system analysis.
25. Buchanan, *Nexus*.
26. See, for example, http://en.wikipedia.org/wiki/Coat_of_arms_of_the_BBC.
27. Friedman, T., *The World Is Flat*, Penguin, New York, 2006.

Index

A

Aboukir Bay, 176
Abstraction, 53, 56, 191
Accidental adversaries, 74, 80
Achilles' heel, 204
Adaptation, 31, 155
Advanced Tools and Methods of Systems Production for Heterogeneous Extensible and Robust Environments, *see* ATMOSPHERE, 96–99
Agile, 139–140, 148, 204
Agility, 10, 107, 111, 151, 156, 159, 160
Airport(s), 197, 203
Aldrich, Ames, 119
Aldrin, Buzz, 95
Alliance, 133, 134, 166
Ancient Greeks, 171
And, 5–7
Anomalous, 188
Ant colonies, 188
Antithesis, 132
Antithetical, 3, 79, 115
Ants, 31, 39, 43, 175
Aperiodicity, 35
Apple Computer, 96
A priori, 7, 38, 91, 147, 149, 152, 156, 161
Arbitration, 143
Archetypes, 69–75, 80–81
Architecture, 42, 49, 59, 62, 102–106, 109–113, 126–130, 136, 148–149, 161–162, 190–200, 203–204, 206
Areas of perspective, 3–10
Aristocratic, 202–204, 206
Armstrong, Neil, 95
Arsonists, xii, 71
Artful Dodger, 12

The Ascent of Man, 35, 42
Ashby, Ross, 36, 159
Assumptions, 8, 12, 39, 47, 120
Asymmetric threat, 8, 124, 131, 206
ATMOSPHERE (Advanced Tools and Methods of Systems Production for Heterogeneous Extensible and Robust Environments), 96–99
Attributes, 1, 34, 146, 207
Automobile, 29, 65, 143, 147, 149–150, 160
 spectrum, 151
Autonomic expertise, 151
Autonomics, 150, 153
Autonomous, 20, 25, 153, 155, 165, 167, 180
 networking, 151
Autonomy, 147, 148, 151, 154, 155–157, 159, 165, 176, 180–181
Axiomatic, 184
Axons, 195

B

Back to biology, 161–163
Bacon, Kevin, 145–146, 190–191
Baer, Robert, 119
Baghdad, 1
Baker-Hamilton report, 2
Balancing loop, 67, 69, 72, 74–75
Bandwidth, 37, 65, 73, 127, 173, 194
Barber, 182
Bar-Yam, Yaneer, 155, 167
Bastille, 177
Battle of Copenhagen, 178
Bayeux Tapestry, 187–188
Bechtolsheim, Andy, 11, 149
Bee Gees, 19

Belong, 23, 25, 31, 33, 48, 134, 147, 151, 153, 157–158, 159, 165, 173, 180, 182
Belonging, 25, 31, 34, 101, 148, 153–154, 155–158, 165, 180–181, 190
Berlin, 46
Biology, xv, 18, 34, 42, 151, 154, 161–163, 167, 206
Bioluminescence, 193
Bishop Odo, 188
Boardmansauser.com, 115
Bobby, 175
Borneo, 194
Bothersome bovines, 181–182
Boundary, xv, 15, 17, 23–24, 26, 29, 30, 31–33, 36, 40, 60, 88, 98, 156, 159, 172–174
Breakfast @ Tiffs 'n' Ease, 79
Brin, Sergey, 11–12, 149
British Empire, 19
Bronowski, Jacob, 35, 42
BRR (business requirements review), 51
Buchanan, Mark, 195, 205–206, 207
Bush, George W., 1–2, 6–7, 45–46, 47, 137–138
Business process architecture, 49, 109, 112
Business requirements review, *see* BRR, 51

C

Calculus, 10, 35, 85, 95, 171
Candy, John, 149
Cape St. Vincent, 177
Capitalism, 82, 205
Carbon, 21, 162
Category, 3, 12, 122, 144
Cats, 7
Causal loop(s), xix, 67, 75, 76–77, 78
CDR (critical design reviews), 51
Central Intelligence Agency, *see* CIA, 118–119, 141
Cerebral cortex, 194
Channel tunnel, 105
Chaords, 35
Chaos, xv, xxi, 34–35, 42, 167, 189, 197, 203, 205, 207
Checkland, Peter, 80–91, 94, 98, 111, 115, 204

Choice, xxi, 2, 4, 6, 10, 14–15, 18, 36, 47, 53, 55, 61, 92, 110, 134, 1 47, 150, 152, 156, 157, 171, 183, 205
Choreography, xvii, xx, 197
Christenson, Clayton, 36
Chrysler, 174
Churches, 33, 75
Churchill, Sir Winston, 20, 42, 53, 175
Churchman, C. West, 16, 33, 42, 204, 207
CIA (Central Intelligence Agency), 118–119, 141
City Slickers, 170
Client-server architecture, 148
Cliques, 118
Cluster, 144, 192–197
Coeducation, 132, 135–136
Coexistence, xx, 132–133
Cohabit, 132–136
Cohabiters, 134
Collingwood, 177
Collisions, 150
Command, 42, 43, 55, 82, 125–128, 139, 151, 154, 158, 160, 166, 167, 174–178, 184, 185, 193–195
Common causes, 117–118
Communications, 36–38, 42, 47, 53, 72, 98, 125–128, 153, 165
Communism, 52, 82, 95
Communities, 4–5, 77, 128, 130, 137, 204
Complementarity, 97
Complex, xviii, xx, 3, 15, 34–36, 39, 46–47, 52, 69, 71, 77, 80, 87, 98, 107–108, 148, 154–155, 159, 160, 162, 163, 167, 187–207
Complexity, xxi, 8, 14–15, 21, 33–36, 42, 53, 80, 91, 105, 167, 188–189, 192, 196–197, 203–205
  theory, 80, 91, 167
Conceptagon, 23–40, 205
Concepts, 19–42, 47, 55, 60, 62, 78, 79–80, 97, 99, 101, 111, 113, 123, 124–126, 129, 131, 138, 167, 170, 172
Conceptual models, xx, 81, 86, 88–94
Conflict, 10, 32, 39–40, 53, 55, 81, 93, 107, 133, 138, 171, 189, 205
Conflicting perspectives, xxi, 172
Connect, 29, 164
Connectivity, 5, 129, 148, 153–159, 165
Conscious integration, 196
Consciousness, 34, 37, 194, 196, 204
Context setting, 196
Contextual tension, 80

Index

Control, 1, 10, 21, 26, 31, 36–43, 46, 51, 58, 64, 66, 86, 108, 115, 1250–128, 139, 141, 148–148, 165–167, 172, 174–178, 183, 201
Conundrum, 22, 35, 181
Cooperation, 14, 48, 74, 120, 132, 134–135, 151
Corbett, Gerald, 84, 87, 102, 105
Counterintuitive, 15, 35, 47, 52, 64, 76, 138, 166, 193
Coworker, 133–135
Cows, 40, 169, 181–182
Creative disobedience, 176–177
Crick, Francis, 161, 167
Crime, 8, 54, 85–86
Crisis, 71
Critical design review, *see* CDR, 51
Cruise, Tom, 144, 146, 173, 189, 190–191
Cuppan, C.D., 155, 166
Currency exchange, 172
Cycle, 16, 33, 68, 69, 126, 152, 201, 204

D

Damon, Matt, 119
Darwin, Charles, 188, 207
Dead wrong, 120
DEC, 96, 149
Democracy, 1, 2, 45, 93–94, 205
Dendrites, 195
de Niro, Robert, 144, 146, 191
Deoxyribonucleic acid, *see* DNA, 133, 161–162
Department of Defense, *see* DoD, 124, 127–128, 130, 147, 166, 167, 199
Determinism, 183, 184
Dialectic, 3
Differentiation, 7, 54, 178–179
Diffusion limited aggregation, *see* DLA, 197, 199, 201
Disaster, 25, 54, 77, 102, 114, 124, 171
Disciples, 99, 135
Disney, Walt, 193
Dispersal, 54
Dissidents, 179
Distribution, 32–34, 74, 155, 169, 191, 199, 202
Diversity, 2, 3, 25, 54, 84, 107, 148, 153–156, 159–160, 165, 178–181
DLA (diffusion limited aggression), 197, 199, 201

DNA (deoxyribonucleic acid), 133, 161–162
Doctrine(s), 2, 61, 131, 141, 154, 176
DoD (Department of Defense), 124, 127–128, 130, 147, 166, 167, 199
Donne, John, 24, 42, 18
Dynamics, 34, 36, 43, 65–78, 80, 134, 150, 157, 175, 207

E

Economic architecture, 102–106
Economic regime, 104
Edward "The Confessor", 187
EEC (European Economic Community), 96–99
Egalitarian, 202–203, 206
Eggs is eggs, 52–53
Einstein, Albert, xvii, xviii, xxi, 43, 52
Eisenhower, Dwight, 82
Electrical grids, 197
Electromagnetic compatibility, 172
Electron, xxi
Emergence (also emergent), xviii, xx, 30–35, 39, 42–43, 47, 71, 92, 97, 99, 147, 149, 151, 154, 155–156, 160–161, 165–167, 79, 188–189, 205
Emotions, 2, 5, 196
Emperor Napoleon, 174
Encapsulation, 53, 105
Engels, Friedrich, 188, 207
Engineering, xvii, xix, xxi, 24, 25, 37–38, 43, 45–64, 72, 80, 84, 86–87, 90–91, 96–97, 100, 102, 109, 129, 159, 169, 162, 166–167, 179, 189
Enigmatic, 189
Enterprise system(s), ix, 106, 166, 179
Entropy, 189
Environment, ix, xv, 23, 31, 36–37, 96, 112, 113, 131, 136, 138, 158, 159, 160–161, 167, 197
Epcot, 193
E Pluribus Unum, 21–22
ESPRIT, 96
Essential characteristics, 146, 157–161
European Economic Community, *see* EEC, 96–99
Evolution, 99–100, 128, 156, 164, 167, 192
Extended enterprise, xx, 29, 47, 102, 105–107, 109, 112, 134–135, 165, 175

Exterior, 23–24, 30–31, 40, 159
Externalities, 172

F

Fairchild Semiconductors, 173
Farmer, 40, 169–170
FBI, 30, 119, 141
A Few Good Men, 146, 164, 173, 189, 190–191
The Fifth Discipline, 65, 74, 78, 115
Firefighter(s), xii, 71, 117
Fireflies, 192–194, 196
Firefly flocks, 197
Fishburne, Lawrence, 175
Fixes that fail, 69–70, 80, 83
Flea, 9–10
Food webs, 190
Football, xi, 5, 7
Forrester, Jay, 43, 65–66
Fort Meade, 119
Fractal, 199
Freedom, 2, 17, 36, 45, 47, 52, 53, 83, 92–93, 95, 146, 157, 159, 175, 177, 205
Free will, xxi, 183
French, 19, 176–177
Friedman, Tom, 53, 64, 206, 207
From prose to picture, 100–102
Function, 17, 22–23, 24–25, 26–30, 148, 157, 161–162, 165, 167, 195
Functional responsibility, 90
Future capabilities, 124–126

G

Gates, 3, 50–51, 61
GE Astrospace, 72
GEC Marconi, 12, 96, 109
Gere, Richard, 22, 57
Getting to the root of the problem, 84–86
Global village, 148
Goals, 8, 45, 68, 80, 96, 117, 123
Goats, 33
Godwinson, Harold, 187
Going off the rails, 102–107
The Good Shepherd, 119
Google, 10–14, 17, 18, 41, 42, 149
Governance, xv, 12, 35, 84, 92, 106, 120, 133, 164, 165
Grand Cherokee Jeep, 174

Growth, 12, 37, 61, 66–68, 70, 73, 102, 104–105, 148, 161, 166, 197, 199, 202–203

H

Haise, Fred, 56
Hanks, Tom, 144, 146, 191
Hanssen, Bob, 119
Harmony, 25, 35–36, 41, 54, 107, 179
Hearts and minds,
Heroics, 71, 135
Heroism, 71
Heuristics, 54, 64
Hierarchy, 23, 30–35, 84, 106, 114, 143–144, 146–149, 151, 166, 167, 174, 177
Hippocampus, 195
Holarchies, 157, 158
Holistic
    approach, 117
    thinking, 99, 205
Holon, 157
Hoover, Edgar, 30
Horizontal integration, 113, 127, 130–132
Hub-and-spoke system, 203
Human
    brain(s), 190, 192, 194–195
    language(s), 190
Hussein, Saddam, 1–2, 15
Hydrogen, 34

I

IBM, 96
Idea, xiii, 4, 12, 21, 22, 39, 53, 58, 81, 132, 175–177, 180, 194
Ideally, this is what we *see*, 86–89
IED (improvised explosive device), 8
If you leave me now, 1–3
Immunity, 172
Immunization, 173
Improvised explosive device, *see* IED, 8
Incarceration, 16, 54, 85
Industrial landscape, 48, 107
Information paralysis, 113, 127, 130,
Infotainment, 47, 143
Injustice, 85
Inputs, xiii, 25–26, 31, 41, 99, 101
Instantiation(s), 5, 161

Integration, xxi, 7, 23, 30, 32, 34, 36, 50, 53–54, 58, 74, 84, 88, 95–97, 106, 111, 113, 121, 123, 126–132, 139–140, 196
Intel, 40, 173
Intel community, 122
Intelligent community, 118–119
Interconnectivity, 26
Interior, 23–24, 30, 40, 106, 143, 148, 159
Internet, 148–149, 158, 159, 188, 190, 197, 199–200, 204
Interoperability, 25, 154–156, 158–159, 166, 183
Interpretation, 26, 33, 100, 112
Into great issues, 95–99
Invasion, 1, 188
Investment backlog, 103
Invisible hand, 107, 152
Iraq, 1–2, 5, 7, 14–16, 93, 94, 119, 120, 205
Iron curtain, 46
It's good to talk, 89–90

J

Jenkins, Ian, xii, 109
Jethro, 143–144
Johnson, Lyndon B., 30
Johnson, Steve, 39, 43
Joint Force Victory, 113, 124
Judgment, xvii, 6, 10, 120, 143, 144, 165, 184, 190
Jumper, John P., 124, 128
Jurisprudence, 183
Just a thought, 19–21
Justice jigsaw, 85

K

Keep it simple stupid, *see* KISS, 36, 159
Kelly, Tom, 56, 64
Kennedy, John Fitzgerald, 45–47, 63, 64, 82, 92, 93, 95
Kennedy, Robert, 95
Kepler, Johannes, 35
Kevin Bacon number, 145, 190–191
Kill chains, 113, 127, 131
Kimmel, Leigh, 176, 185
KISS (keep it simple stupid), 36, 159
Kranz, Gene, 56

L

Latent, 134
Leadership, xviii, xxii, 39, 77, 98, 120–121, 130, 143, 178–179
Legacy, 69, 80, 125, 129, 131, 155–157, 159
Leibniz, Gottfried, 35, 171
Less auto more mobile, 149–151
Life cycle, 48–50, 55, 61, 112, 162
Life's rich tapestry, 187–190
Limits to growth, 70, 73, 202
Linear thinking, 99
Links, 12–13, 55, 67, 87, 97, 101, 111, 134, 144, 164, 191, 192–195, 199–200, 203–205, 206
Lockheed Corporation, 173
Lockheed Martin, 72
The Long haul, 58–61
The Long unwinding road map, 91
Loops, 26, 38, 68, 70, 72, 78
Lord Spencer, 178
Louis XVI, 177
Lovell, James, 55–56, 64

M

The Madder of all wars, 124–132
Mainframe, 148
Maintainability, 60
Make my joy complete, 169–170
Making sense of togetherness, 122–124, 130–132, 151
Malaysia, 194
Mars, 146, 191, 201
Martin, Steve, 149
Matra Marconi Space, 72
Mattingly, Ken, 56
Mavericks, 179
Meme, 175
Memory, 4, 6, 148, 194–196
Meta-languages, 183
Metaphor, 61, 80, 173, 206
Methodology, xix, xx, 54–55, 58, 80–81, 83–84, 86, 88, 91, 111, 114, 204
Minicomputer, 96, 148
Miracle(s), 143, 173
Mississippi River, 3, 197–198
Modeling and simulation, 55–58, 61, 128, 165
Models of
 ideality, xx, 87
 reality, 66, 86
Moore, Demi, 190

Moses, 64, 109, 143–145, 146–147
Movement detection, 196
MRI scans, 195

N

NASA (National Aeronautics and Space Administration), xv, 82–83, 207
National Aeronautics and Space Administration, *see* NASA, xv, 82–83, 207
National Security Agency (NSA), 118–119
Negative feedback, 67
Negroponte, John, 118
Neighbor(s), xvii, 4, 26, 31, 39–40, 107, 111, 152, 169–170, 175, 182, 192–196
Nelson, Admiral Horatio Lord, 176
Net-centricity, 140, 158
Network
  evolution, 164, 192
  of networks, 127, 131
  rail, 103
Neumann, John Von, 36
Neural networks, 197
Neurological science, 194
Neuron(s), xvii, 192–193, 195–196, 206
New
  constellation, 126–130
  Guinea, 192–194
  product development, 48, 56, 71
Newton, Sir Isaac, 20, 35, 171
Nicholson, Jack, 144, 146, 190–191
Nihilistic, 33
Nile, 176–177
Nitrogen bases, 161
Nobel Prize, 160
Node(s), 67, 97, 101, 113, 128–129, 131, 139–140, 144, 148, 164, 191, 199–200, 205, 206
Nokia, 53
Norfolk, 177
Normal service will never be resumed, 107–111
Not, 7–9
NSA, (National Security Agency), 118–119
Nugatory, xx, 6, 48, 52, 83, 171

O

Obviates, 160, 174
Ockham's Razor, 36
Of, 143–166
Oil, 2, 75–76, 205
Omnipotence, 183, 185
Ontologies, 183
Opaque, 183
Openness, 30–35, 128, 151
Opraf, 89, 103, 106
Or, 3–5
Oracle of bacon, 190
Ordered liberty, 176
Ordered questioning, 90
Order forms, 146–147
Outputs, 25–26, 29, 31, 39, 41, 99, 101, 105

P

P&G, (Proctor and Gamble), 74
Pacino, Al, 144
Page, Larry, 11, 149
Papua New Guinea, 192, 193
Paradigm, 15, 24, 66, 96, 131, 143, 147, 164
Paradox, xxi, 3, 9–10, 13, 16, 17, 52, 54, 61, 63, 80, 122, 133, 162, 169–185, 188–190, 192, 196, 197, 203–205
Paralysis by analysis, 52, 57, 83, 91, 121
Parsimony, 35–36, 37, 41, 107, 179
Part, 58, 129
Passing through, 50–52
Pattern recognition, 196
Pattern(s), 17, 33–34, 39, 55, 65–66, 68–69, 151–152, 158–159, 162, 179, 189, 194–196, 198–200, 203, 205
Paul II, Pope John, 191
PDR (preliminary design review), 51, 61
Peer-to-peer, 113, 127, 131
Pentagon, 118
Perception, 15–16, 32–33, 204
Performance regimes, 103
Perspective(s), ix, xi, xv, xviii, xx, xxi, 1–18, 50, 54, 56–57, 60–61, 80, 86, 91, 105–108, 112, 114, 141, 143, 152, 159, 161, 166, 167, 170, 172, 187, 205
  on Google, 10–14
Phenomena, xviii, xxi, 33–34, 42, 46, 79, 154, 155
Phosphates, 161
Phrenology, 194
Pilgrim's Progress, 50
Pioneers, 80
Pioneering, 92, 207
Political integration, 96

Index

Pollack, Kevin, 173, 190
Positive feedback, 67
Postlude, 91, 111–113, 204–205
Power law, 198–201
Preliminary design review, *see* PDR, 51, 61
Pressed into action, 45–47
Pretty Woman, 22, 57
The Price and prize of togetherness, 151–154
Priests, 33
Principles for togetherness, 132–136
Probability, 34, 145, 200
Problematique, xix, 47, 83, 90–91, 121, 155, 157–160
Process, 26–30, 76, 114, 127, 133, 139
Proctor and Gamble, *see* P&G, 74
Production readiness review, *see* PRR, 51
Progeny, 148
Proteins, 190
PRR, (production readiness review), 51
Pythagoras, 35

Q

Queen, 37, 39, 78, 201, 207
Quite another story, 61–63

R

Rail accidents, 103
Railtrack, 84, 89, 102–106
Random
  links, 192–193
  networks, 192, 204
  rewiring, 193
Ready, fire, aim, 193–197
Reagan, Ronald, 45–47
Reciprocal loyalty, 176–177
Reconstruct the past, 90
Redford, Robert, 47
Reductionism, 25, 33, 35, 42, 162
Regular networks, 192
Rehabilitation, 16, 85
Reinforcing loop, 67, 69, 72, 74, 201
Relationships, xix, xx, 5, 9, 23, 24–25, 29, 32–34, 36, 38–39, 41, 47, 66–67, 76, 93, 97, 99, 101, 105, 113, 145–147, 158, 160, 162–163, 172, 180, 189, 192, 206
Reliability, 37, 60, 102, 203
Reporting structures, 90

Reproduction, 161
Requirements, 27–28, 51, 52–54, 57, 60–62, 73, 76, 86, 100, 102, 109, 111–113, 128, 150
Requisite variety, 36, 38, 159, 179
Resilience, 10, 106–107, 111, 165, 196, 203, 204
Returning to Iraq, 14–16
Reverse
  engineered, 90
  engineering, 102
Review, 8, 27, 33, 51–52, 55, 61, 64, 72, 99, 105, 114, 128, 138
Rivers, 3, 197, 199
River systems, 3, 198–199
Robb-Silberman, 120
Roberts, Julia, 22, 57, 144, 146, 191
Roche, James G., 124
Rolling stock
  company, *see* ROSCo, 89, 103, 106
  investment, 105
Rolls-Royce, xii, 48–49, 70–71, 165
Root definition(s), 81, 84–88, 90, 92
ROSCo (Rolling Stock Company), 89, 103, 106
Rosin, Hanna, 118–119, 120
Rules, 34–35, 37, 39, 54, 77, 100–101, 111, 113, 128–129, 147, 199–200, 202
Russell, Bertrand, 182

S

Sacrifice, 2, 10, 41, 180
Safety, 20, 25, 105, 143, 150, 203
Sage, Andrew, 155, 166
Samarra, 2
The Sameness of systems, 22–23
Samos. 35
Sandoz, David, xi, 39
Satellite, 72, 82, 108, 128, 165, 201
  navigation, 143
Science of networks, 80, 192
Scientific method, 33
Scrambled eggs, 79
SDI (Strategic Defense Initiative), 46
Security, 37, 77, 106, 111, 118–119, 125, 130, 136, 138, 140, 141, 151, 189, 203
See what I mean, 53–55
Self-
  achievement, 178
  control, 165, 175–176
  regulation, 103
  reinforcing, 199

similar, 199
Semantic interoperability, 183
Senge, Peter, 38, 43, 65–66, 74, 78, 80, 115, 201, 202, 204
Servers, 197
  and clients, 66–69
The Shawshank Redemption, 85
Sheep, 7
Shia, 1–2
Shifting the burden, 71–72, 80
Short- and long-term goals, 80
Sides of bacon, 190–192
Significant others, 203–204
Simon, Herbert, 160, 167
Simultaneously tenable viewpoints, 79–80, 105, 133
Simultaneous opposites, 172, 180
Sinese, Garry, 56
Skunk works, 173
Slavery, 41, 143
Slime molds, 188
Smith, Adam, 188, 207
Snowballs and *see*saws, 197–203
Soccer, 5, 179
Social
  circles, 190, 197
  networks, 144–146, 147, 195
  systems, 91
Soft, 79–91
  systems methodology, xx, 55, 82, 84, 88, 111, 114, 204
Softly, as I lead you, 80–81
Solomon, King, 10, 184
Solutions creep, 131
Soviet Union, 46, 82, 92
Space race, 82
Spoiled for choice, 55
Sponsored links, 12–13
Sputnik, 45, 82–83
SRR (system requirements review), 51
Stakeholder(s), xxi, 6, 7, 50, 54, 61–62, 75, 84, 89–90, 91, 105, 107, 141, 155, 170, 179
  tension, 81
Stallkamp, Thomas, 174–175
Stampede, 1
Stanford University, 11
Steiger, Rod, 191
The Story so far, The, xvii–xxii
Stovepipe, 121, 123, 130–132
Strategic Defense Initiative, *see* SDI, 46
Strategic intent, 55, 97–101, 111–112, 120
Strategic Rail Authority, 104

Strategy bridge, 100
Streisand, Barbara, 47
Strogatz, Steve, 192, 194, 196, 202, 207
Structure, xiii, xx, 12, 22, 26–30, 38, 42, 49, 57, 62, 66, 73–74, 87, 90, 97, 101, 133, 150, 152, 17, 161–163, 164, 167, 172, 190, 203
Subsidiarity, 96
Sugars, 161
Sun Microsystems, 148
Supply chain, 134
Supportability, 59–60
Surfers, 11, 197
Surrender, 19, 180
Suspension system, 150
Sutherland, Kiefer, 146, 190–191
Swanick, Brian, xi, 39
Swigert, Jack, 56
Symptoms, 69, 82–83
Synchronicity, 192–193, 196
System
  paradoxes, 172–181
  requirements review, *see* SRR, 51
  under consideration, 163
  under observation, 23
Systemic, xviii, xix, 25, 101, 105, 115, 117, 122, 133
  diagram, xx, 95, 97
  failure, 42, 46, 117
  models, xx, 89
  structure, 66, 73
Systemigram, xx, 87–89, 95–115, 117, 121–124, 126, 129–131, 138, 140–141, 205
System of
  interest, 23
  systems (SoS), xx, 25, 32, 39, 46, 47, 87, 112, 124, 131, 154–161, 165, 166–167
SystemiTool, 115
Systems
  DNA, 161–162
  integration, 97
  language, 65–66, 69, 73, 80, 91
  practice, xxi, 39, 60, 63, 80, 88, 98, 99, 115
  thinking, xviii, xix, xx, xxi, xxii, 2, 11, 12, 16, 17, 30, 38–39, 42, 47, 52, 54–55, 57, 61, 63, 65, 66, 68–69, 77, 78, 79–81, 86–89, 91, 97–99, 105, 113, 115, 117, 121–122, 154, 163, 170–172, 189, 192, 204

Index

T

Tapestry, 187–189, 207
   of tension, 80
Technology
   integration, 96
   networks, 147–149
   society, 150
Temporal tension, 80
Ten Commandments, 36
Tension, 25, 36, 51–52, 58, 64, 80–81, 171, 174–176, 179, 181, 183, 184
Terror, 2, 42, 46
Test requirements review, *see* TRR, 51
Thailand, 194
Theater tickets, 150
Thinks can only get better, 65–78
Thought foci, 5–6, 8
Time to think, 16–18, 40–42, 63–64, 75–78, 91–94, 113–141, 163–166, 182–184, 205–207
Togetherness, xvii, xix, xx, xxi, 54, 117–141, 144, 146, 151–154, 161–163, 165, 178–180, 206
Tools, xix, 2, 19, 54, 56, 59, 80, 96, 139–140, 205
Topology, 158, 164
Trade-off studies, 4, 60, 73
Trafalgar, 177–178
Traffic jams, 42, 150, 160
Tragedy of the commons, 72–73, 80
Train-operating companies, 103
Transformation(s), 25–26, 29, 41, 55, 64, 99, 101, 121, 124, 126, 128, 130–131, 139, 157
Transmission system, 150
Transparent, 183
Triad, 25–26, 29, 36–37
Trivial many, 203
Troubridge, 176
TRR (test requirements review), 51
Turing, Alan, 188, 207
Turner, Tina, 36

U

Uncertainty, 34, 197, 202
Unintended consequences, 69, 75, 89, 91
Unity, 23, 35–36, 169
University of crime, 54
U.S. soldiers, 1–2

V

Variety, 7, 8, 21–22, 25, 32, 35–36, 38, 41, 53–54, 57, 63, 91, 106, 112, 133, 145, 148, 159–160, 171, 179, 202
Viewpoint, 54, 110, 133, 170
Vise, David, 11, 18
Visual cortex, 195
Vital few, 203–204

W

Waldrop, Mitchell, 21, 42, 167, 188, 207
Wal-Mart, 74–75
Wannabe, 144, 147
Washington, George, 174
Washington Post, 118
Watson, James, 161, 167
Watts, Duncan, 192, 194, 196, 202, 207
The Way
   it is, 118–120
   it must be, OK?, 120–122
   we were, 47–48
A Weakness stronger than strength, 192–193
Weapons of Mass Destruction, *see* WMD, 120
Weinberg, Gerry, 22, 42
Wellspring, 47, 80, 171
Westminster Abbey, 187
What *seems* to be the problem?, 81–84
What's the Big Idea, 21–23
Whole(s), xix, xx–xxii, 9, 24–26, 29, 31, 34, 39, 42, 47, 54, 58, 61, 66, 84, 99, 106, 117, 132, 146–147, 157–158, 162, 167, 171–172, 180–181, 188–189, 196, 204–205
William, Duke of Normandy, 187
William of Ockham, 36
Wisdom, 3, 5, 9–10, 12–13, 34, 51, 55, 63, 95, 122, 151, 160, 170–171, 182, 184, 204–205
   of crowds, 42, 151, 166
Witan, 187
WMD (Weapons of Mass Destruction), 120
The World of both, 170–172
World wide web, 11, 148, 158, 190, 192

Z

Zeno's paradoxes, 171